AF291298

Plant Organelle DNA Maintenance

Plant Organelle DNA Maintenance

Editors

Brent L. Nielsen
Niaz Ahmad

MDPI • Basel • Beijing • Wuhan • Barcelona • Belgrade • Manchester • Tokyo • Cluj • Tianjin

Editors
Brent L. Nielsen
Brigham Young University
USA

Niaz Ahmad
National Institute for Biotechnology & Genetic Engineering (NIBGE)
Pakistan

Editorial Office
MDPI
St. Alban-Anlage 66
4052 Basel, Switzerland

This is a reprint of articles from the Special Issue published online in the open access journal *Plants* (ISSN 2223-7747) (available at: https://www.mdpi.com/journal/plants/special_issues/plant_organelle_dna).

For citation purposes, cite each article independently as indicated on the article page online and as indicated below:

LastName, A.A.; LastName, B.B.; LastName, C.C. Article Title. *Journal Name* **Year**, *Article Number*, Page Range.

ISBN 978-3-03936-692-7 (Hbk)
ISBN 978-3-03936-693-4 (PDF)

Contents

About the Editors

Brent L. Nielsen, Professor of Microbiology and Molecular Biology.
Brigham Young University, Provo, Utah 84602 USA.
B.S., Microbiology—Brigham Young University, 1980.
Ph.D., Microbiology—Oregon State University, 1985.
Postdoctoral Research, Molecular Biology and Biochemistry—University of California, Irvine, 1985–1988.
Assistant, Associate, and Full Professor—Auburn University, Alabama, 1988–2000 Professor—Brigham Young University, 2000–present.
Chair—Department of Microbiology and Molecular Biology, Brigham Young University, 2005–2011.
Research Interests: plant molecular biology; DNA replication and recombination; mechanisms of salt tolerance in plants; salt-tolerant plant growth-promoting bacteria.

Niaz Ahmad, Ph.D.
Senior Scientist.
Agricultural Biotechnology Division, National Institute for Biotechnology & Genetic Engineering (NIBGE), Jhang Road, Faisalabad 38000, Pakistan.
MPhil—Quaid-i-Azam University, Islamabad, Pakistan, 2005.
Ph.D.—Imperial College London, United Kingdom, 2012.
Assistant Professor—NIBGE, Jhang Road, Faisalabad, Pakistan, 2012–2013.
Senior Scientist—NIBGE, Jhang Road, Faisalabad, Pakistan, 2013–present.

Preface to "Plant Organelle DNA Maintenance"

Plastids and mitochondria contain their own genomes that encode genes required for function of these organelles. These genomes vary greatly in size. Organelle DNA is found at high copy numbers per mitochondria or chloroplast, and this number varies greatly in different tissues during plant development. It is not yet clear how the genome replication, repair, and overall maintenance activities of mitochondria and chloroplasts are coordinated in response to cellular and environmental cues. The relationships between genome copy number variation and the mechanism(s) by which the genomes are maintained through different developmental stages are yet to be fully understood. This Special Issue comprises several contributions that address the current knowledge of higher plant organelle genome sequence and structural analysis, along with reports about DNA replication, recombination, and DNA repair. We greatly appreciate the efforts of those who contributed to this Special Issue with manuscripts or reviews.

Brent L. Nielsen, Niaz Ahmad
Editors

Editorial

Plant Organelle DNA Maintenance

Niaz Ahmad [1] and Brent L. Nielsen [2,*]

[1] Agricultural Biotechnology Division, National Institute for Biotechnology & Genetic Engineering, Jhang Road, Faisalabad 38000, Pakistan; niazbloch@yahoo.com
[2] Department of Microbiology & Molecular Biology, Brigham Young University, Provo, UT 84602, USA
* Correspondence: brentnielsen@byu.edu; Tel.: +1-801-422-1102

Received: 12 May 2020; Accepted: 21 May 2020; Published: 28 May 2020

Abstract: Plant cells contain two double membrane bound organelles, plastids and mitochondria, that contain their own genomes. There is a very large variation in the sizes of mitochondrial genomes in higher plants, while the plastid genome remains relatively uniform across different species. One of the curious features of the organelle DNA is that it exists in a high copy number per mitochondria or chloroplast, which varies greatly in different tissues during plant development. The variations in copy number, morphology and genomic content reflect the diversity in organelle functions. The link between the metabolic needs of a cell and the capacity of mitochondria and chloroplasts to fulfill this demand is thought to act as a selective force on the number of organelles and genome copies per organelle. However, it is not yet clear how the activities of mitochondria and chloroplasts are coordinated in response to cellular and environmental cues. The relationship between genome copy number variation and the mechanism(s) by which the genomes are maintained through different developmental stages are yet to be fully understood. This Special Issue has several contributions that address current knowledge of higher plant organelle DNA. Here we briefly introduce these articles that discuss the importance of different aspects of the organelle genome in higher plants.

Keywords: organelles; plastid phylogenetics; DNA replication; DNA recombination; plant organelle genome structure

---

1. Introduction

Mitochondrial and chloroplast genomes have been studied for nearly 60 years, and much is known about the structure, gene content, expression and other characteristics of organelle genomes. Plant mitochondria, in great contrast to mitochondria in other organisms, have a very wide range in genome size and appear to be predominantly linear in structure. Chloroplast genomes are generally more consistent in size and may exist in both circular and linear forms [1]. Major questions that have persisted, however, relate to how these plant organelle genomes are replicated, repaired and maintained.

The endosymbiotic origin of mitochondria and chloroplasts has long been accepted, suggesting that the DNA replication mechanisms would be similar to their bacterial ancestors. However, none of the DNA replication, DNA recombination or DNA repair proteins required for maintenance of the organelle genomes are encoded in either of these organelle genomes [1,2]. The genes for these proteins are nuclear encoded and the protein products are imported into the proper organelle for function. The analysis of the minimal replisome in plant mitochondria and chloroplasts to date indicates that it is phage-like, most similar to the bacteriophage T7 system [1,2].

Plant organelle genomes—both mitochondrial and plastid—are relatively small (though larger than animal mitochondrial genomes) and are found in variable and sometimes incredibly high copy numbers. Their copy number varies from tissue to tissue, depending mainly on the age and type of the tissue. While higher plant mitochondria show a considerable degree of variation in their genome sizes, plastid genome size remains fairly uniform among different species [1].

Organellar genome substitution rates are much lower than the nuclear genome due to homologous recombination, which allows these molecules to be useful as molecular clocks to carry out evolutionary studies and determine genetic relationships among different taxa. In addition to phylogenetic studies, mitochondrial genomes provide excellent tools to gain insights into cytoplasmic male sterility (CMS) sources in plants.

Organelle genome stability is an absolute requirement for normal cell growth and function, and for proliferation. Faithful replication and repair of organellar genomes are therefore necessary to avoid genome instability, which may result in detrimental phenotypes. The mechanisms by which the organelle genomes in different tissues are replicated, maintained and repaired are not fully understood. This Special Issue was hosted to address unanswered questions related to organelle genome maintenance in plants. The Special Issue has eight helpful contributions that have attempted to answer some of these questions. The first four articles deal with questions regarding the maintenance and replication of the organelle genomes, whilst the last four articles deal with organellar genome sequencing and structural analyses.

2. Key Messages

2.1. Mechanisms for Plant Organelle Genome Replication and Repair

Brieba [2] summarizes the current literature to point out the importance of homologous DNA recombination in maintaining the integrity of plant mitochondrial genomes. Plant mitochondrial DNA (mtDNA) appears to be predominantly linear in nature and lacks any clearly identified replication origins. DNA replication proteins functional in both mitochondria and chloroplasts have been identified, and surprisingly, there are multiple copies of the genes for most of these proteins [1,2]. Other evidence in the literature suggests the presence of more than one mechanism for the replication of each of the plant organelle genomes [1,2]. These possibilities include typical primer-driven replication from distinct origins, recombination-dependent replication (RDR) and rolling circle replication.

Each of these mechanisms requires several common replication functions, but it is unclear how priming of DNA replication occurs in plant organelles. Twinkle is a DNA primase/helicase that is localized to both organelles in plants. In animal cells, this protein is absolutely required, but plant mutants are fully viable with no apparent differences compared with wild-type plants [1,2]. These two contributions discuss these observations and the possibility of other proteins being responsible for priming in plant organelles.

Plant organelle genomes, in particular those in mitochondria, contain repeat sequences that facilitate intramolecular recombination to generate various sub-genomic molecules. Recombination across these repeats, however, is regulated and suppressed by surveillance mechanisms including MSH1 and some Rec proteins [2,3]. MSH1 mutants in *Psychomytrella* show increased recombination across intermediate sized repeats relative to the wild-type [3].

Mutations in plant mitochondrial gene coding regions are rare, suggesting efficient repair of miss-paired or damaged bases such as uracil generated by deamination of cytidine. Proteins for mismatch and nucleotide excision repair have not been found in plant mitochondria [4], but base excision and double-strand break repair appear to be functional in both plant organelles. In Wynn et al. [4], uracil N-glycosylase (UNG) mutants were examined to determine if this defect in base excision repair (BER) would lead to more mutations, but this was not found. They did find that several genes that encode proteins involved in double-strand break repair (DSBR) were upregulated in the UNG mutants. This led to the conclusion that most damage to plant mitochondrial DNA is repaired by DSBR with further contribution by BER [4]. This implication of a major role for DSBR, which utilizes homologous DNA recombination, is supported by many studies that report the adverse effects of mutations in DNA recombination proteins on plant organelle integrity [1,3].

2.2. Characteristics and Rearrangements of Plant Organelle Genome Structure

Our knowledge about the genetic diversity of plant organelle genomes is currently based on small intergenic polymorphism regions. These same regions are used to conduct evolutionary studies and construct phylogenetic trees. However, internal transcribed spacer (ITS)-based trees do not have power to resolve the phylogenetic relationships among species and groups. With advances in sequencing technologies, organellar genomics is now moving away from individual gene analysis to whole genomes for the reconstruction of phylogenies at lower taxonomic levels, to study divergence among different groups, and for the structural analysis of these genomes. The availability of complete organellar genome sequences allows the construction of "species trees'" by which the accuracy of the phylogenetic relationships is greatly increased. Although gene-based phylogenetic trees, commonly referred to as "gene trees" could be readily obtained previously, they fail to trace all of the evolutionary events since not all genes evolve in a similar manner in the same lineage. Whole genome sequencing (WGS) provides valuable information about the occurrence of nucleotide recombination events in the organelle genomes.

The genus *Mammillaria* is an important genus due to its highly rich genetic diversity. Several of its species have been included in the Red List of Threatened Species of the International Union for Conservation of Nature. However, molecular marker-based phylogenetic studies fail to resolve the relationships among species of the *Mammillaria* genus. Solórzano et al. [5] obtained the complete chloroplast genome (plastome) sequences of seven different species of the short-globose cacti of *Mammillaria* and used a species tree to resolve the species of this genus.

The genus *Lilium* is widely distributed in the cold and temperate regions of the Northern Hemisphere. Many attempts using molecular phylogenetic approaches have been made; however, there remain unresolved clades in the genus. Kim et al. [6] sequenced the chloroplast genome of 28 *Lilium* species and used these sequences to construct a species tree in order to increase the accuracy of phylogenetic relationships for unresolved taxa.

WGS data also facilitate studies on the mechanisms of gene loss, recombination events, transfer of genes to the nucleus and evolution of different traits in different lineages. Cytoplasmic male sterility (CMS) has been extensively studied in sunflower (*Helianthus annuus* L.), which has more than 70 sources. This earlier work was primarily done by restriction fragment length polymorphism (RFLP) analysis, which does not provide a detailed explanation of the underlying events. One of the spontaneously occurring CMS sources in sunflower is ANN2, which is quite complicated and cannot be restored completely. Makarenko et al. [7] used the WGS method to gain insights into the structural reorganization of the mitochondrial genome that generates ANN2 in sunflower. The authors compared the mitochondrial genome sequences from a male sterile line with that of a male-fertile line and observed several reorganization events (deletions and insertions) and several new transcriptionally active open reading frames (ORFs) in the mitochondrial genome of the male sterile line. The deletion events resulted in the partial removal of the *atp6* (*orf1197*) gene and complete elimination of *orf777*, indicating a major role of the *atp6* gene in ANN2-type CMS.

It is now well-established that chloroplast genomes evolved from free-living cyanobacterial cells. The endosymbiotic events resulting in the conversion of these cells into organelles were accompanied by a massive transfer of their genetic material to the host cell nucleus, with only a small set of essential genes retained in their genomes. In heterotrophic plant species, with the loss of selection pressure over photosynthesis-related genes, this genome shrinkage is even higher. The plastome of such plant species provides an excellent tool to study the loss of organelle genetic material to the nucleus while they shift from the autotrophic to the heterotrophic mode. Jost et al. [8] sequenced the plastome of a holoparasitic plant species, *Prosopanche americana*, which has only 24 housekeeping genes and has lost the inverted repeats (IR) that are thought to stabilize the genome. The genetic machinery of *P. americana* shows an incredibly high degree of bias towards "AT"-rich codons, with >90% of the plastome sequence containing "A" and "T" nucleotides.

3. Future Directions

Considerable progress has been made on sequencing and characterizing plant organelle genomes since they were discovered about sixty years ago. However, there are numerous aspects of their replication and maintenance that are still unclear. Further work is needed to determine whether there is a single mechanism for DNA replication in plant mitochondria and chloroplasts. If there is more than one mechanism, what are they and when/how do they function? Do these mechanisms function at different stages of plant growth or in response to specific signals? It is still unclear what mechanisms control the initiation of plant organelle genome replication and copy number. What determines and regulates genome copy number in meristem and mature plant tissues? Why does genome copy number vary so widely in different tissues and ages of plants? These are important questions that require further research.

While DNA replication origins have been characterized in chloroplasts, there is as yet no conclusive evidence for distinct replication origins in plant mitochondria. This is a challenge that has been studied but no clear conclusions have been published. New approaches are needed to address this question. Along with this, it is unclear how plant organelle DNA replication is primed. Since Twinkle mutants are viable, there must be a separate mechanism for generating primers unless they are not needed (i.e., if homologous strand invasion provides the needed 3′ ends for DNA synthesis). If this is the case, this would suggest that homologous DNA recombination is a major contributor to genome replication in addition to its likely role in DNA repair. The role of DNA recombination in the replication and maintenance of plant organelle genomes thus deserves further study.

Regarding repair of DNA damage, the current literature suggests that double-stranded break repair and base excision repair may be the only mechanisms that function in plant organelles. If so, what are the relative contributions of each process in these organelles? Does nonhomologous end joining occur in plant organelles? And how does it affect genome integrity?

Finally, there are still questions regarding the evolution and variation of plant organelle genomes. Evolutionary adaptations may play a central role in plant mitochondrial genome size variation. However, it is unclear what mechanisms and selection processes determine plant organelle genome size. In addition, the mechanisms involved in purifying the selection by which parasitic plants keep some house-keeping genes, while losing other genes, and the IRs are unclear. Does this reduction in content affect genome stability? This is a promising area of research, and we look forward to new studies in the future to address these many questions.

Funding: This research received no external funding.

Acknowledgments: We thank our colleagues who contributed articles to this Special Issue.

Conflicts of Interest: The authors declare no conflict of interest.

References

1. Morley, S.; Ahmad, N.; Nielsen, B. Plant Organelle Genome Replication. *Plants* **2019**, *8*, 358. [CrossRef]
2. Brieba, L. Structure—Function Analysis Reveals the Singularity of Plant Mitochondrial DNA Replication Components: A Mosaic and Redundant System. *Plants* **2019**, *8*, 533. [CrossRef] [PubMed]
3. Odahara, M. Factors Affecting Organelle Genome Stability in *Physcomitrella* patens. *Plants* **2020**, *9*, 145. [CrossRef] [PubMed]
4. Wynn, E.; Purfeerst, E.; Christensen, A. Mitochondrial DNA Repair in an *Arabidopsis thaliana* Uracil N-Glycosylase Mutant. *Plants* **2020**, *9*, 261. [CrossRef] [PubMed]
5. Solórzano, S.; Chincoya, D.; Sanchez-Flores, A.; Estrada, K.; Díaz-Velásquez, C.; González-Rodríguez, A.; Vaca-Paniagua, F.; Dávila, P.; Arias, S. De Novo Assembly Discovered Novel Structures in Genome of Plastids and Revealed Divergent Inverted Repeats in Mammillaria (*Cactaceae*, Caryophyllales). *Plants* **2019**, *8*, 392. [CrossRef] [PubMed]
6. Kim, H.; Lim, K.; Kim, J. New Insights on Lilium Phylogeny Based on a Comparative Phylogenomic Study Using Complete Plastome Sequences. *Plants* **2019**, *8*, 547. [CrossRef] [PubMed]

7. Makarenko, M.; Usatov, A.; Tatarinova, T.; Azarin, K.; Logacheva, M.; Gavrilova, V.; Kornienko, I.; Horn, R. Organization Features of the Mitochondrial Genome of Sunflower (*Helianthus annuus* L.) with ANN2-Type Male-Sterile Cytoplasm. *Plants* **2019**, *8*, 439. [CrossRef] [PubMed]

8. Jost, M.; Naumann, J.; Rocamundi, N.; Cocucci, A.; Wanke, S. The First Plastid Genome of the Holoparasitic Genus Prosopanche (*Hydnoraceae*). *Plants* **2020**, *9*, 306. [CrossRef] [PubMed]

Review

Plant Organelle Genome Replication

Stewart A. Morley [1,†], Niaz Ahmad [2] and Brent L. Nielsen [1,*]

[1] Department of Microbiology & Molecular Biology, Brigham Young University, Provo, UT 84602, USA; stewart.morley@usda.gov
[2] Agricultural Biotechnology Division, National Institute for Biotechnology & Genetic Engineering, Jhang Road, Faisalabad, Punjab 44000, Pakistan; niazbloch@yahoo.com
* Correspondence: brentnielsen@byu.edu; Tel.: +1-801-422-1102
† Current address: United States Department of Agriculture, 975 Warson Rd, St. Louis, Mo 63132, USA.

Received: 13 August 2019; Accepted: 18 September 2019; Published: 21 September 2019

Abstract: Mitochondria and chloroplasts perform essential functions in respiration, ATP production, and photosynthesis, and both organelles contain genomes that encode only some of the proteins that are required for these functions. The proteins and mechanisms for organelle DNA replication are very similar to bacterial or phage systems. The minimal replisome may consist of DNA polymerase, a primase/helicase, and a single-stranded DNA binding protein (SSB), similar to that found in bacteriophage T7. In *Arabidopsis*, there are two genes for organellar DNA polymerases and multiple potential genes for SSB, but there is only one known primase/helicase protein to date. Genome copy number varies widely between type and age of plant tissues. Replication mechanisms are only poorly understood at present, and may involve multiple processes, including recombination-dependent replication (RDR) in plant mitochondria and perhaps also in chloroplasts. There are still important questions remaining as to how the genomes are maintained in new organelles, and how genome copy number is determined. This review summarizes our current understanding of these processes.

Keywords: DNA replication; recombination-dependent replication (RDR); plant mitochondrial DNA; chloroplast DNA; DNA repair

---

1. Introduction

1.1. Discovery of Mitochondria and Chloroplasts

In 1665 Robert Hooke became the first person to observe cells with a simple microscope [1]. Almost one hundred and fifty years later in 1804 Franz Bauer described the discovery of the first observed organelle, the nucleus [2]. In 1890 Richard Altmann described what he called "bio-blasts," or what we now call mitochondria [3]. Around the same time in 1883, A. F. W. Schimper described "chloroplastids", what we now know as chloroplasts [4]. Both of these organelles house important biochemical reactions that are essential for cell survival; mitochondria generate ATP, and chloroplasts are the site of photosynthesis, and both house other important functions [3,5].

1.2. Evolutionary Origins of Each Organelle

Both mitochondria and chloroplasts are believed to have originated through endosymbiosis. Free-living aerobic α-proteobacterium-like cells taken up by a nucleus-containing (but amitochondriate) host cell gradually developed into mitochondria. Chloroplasts are also thought to have developed by a similar process, in which a eukaryotic cell (containing the mitochondria) engulfed a photosynthesizing prokaryotic cell, which eventually evolved into present-day chloroplasts. Over the course of evolution, both of these incoming cells entered into an endosymbiotic relationship with the host, synchronizing their own division with that of the host cell, transferring their genetic material to the host nucleus, and becoming permanent cellular structures of exogenous origin [5–7]. Despite the extensive knowledge of organelle genome structure, evolution and content that has been reported, there are still important questions regarding how these genomes are replicated, and the proteins and control mechanisms involved.

2. Organelle Genomes and Structure

2.1. Genome Size

Endosymbiosis is accompanied with massive gene transfer to the nucleus of the host cell, resulting in considerable size reduction of the genome of the incoming cells. This is observed in the considerable size reduction in the mitochondrial and chloroplast genomes and the presence of mitochondrial and chloroplast DNA (cpDNA) sequences in the nuclear genomes of many plant species [5,8,9]. Mitochondrial genomes in plants have evolved very differently as compared to animal mitochondrial genomes (Figure 1). Most animal mitochondrial genomes are roughly 16 kb in size [10], and the number of genome copies per mitochondrion varies from study to study. Older estimates place as many as 10 copies per organelle [11], whereas more recent data suggests it may be as low as one [12]. Regardless of the actual number, mitochondrial genome copy number is thought to be tightly regulated in animal cells [13]. In contrast, plant mitochondrial genomes are much larger, and have tremendous size variations (187–2400 kb [14]) among different species. In addition, significant diversity in the number of copies of mitochondrial DNA (mtDNA) per organelle have been reported in different species, tissues and cell types [15,16]. The reasons for these copy number differences are unclear.

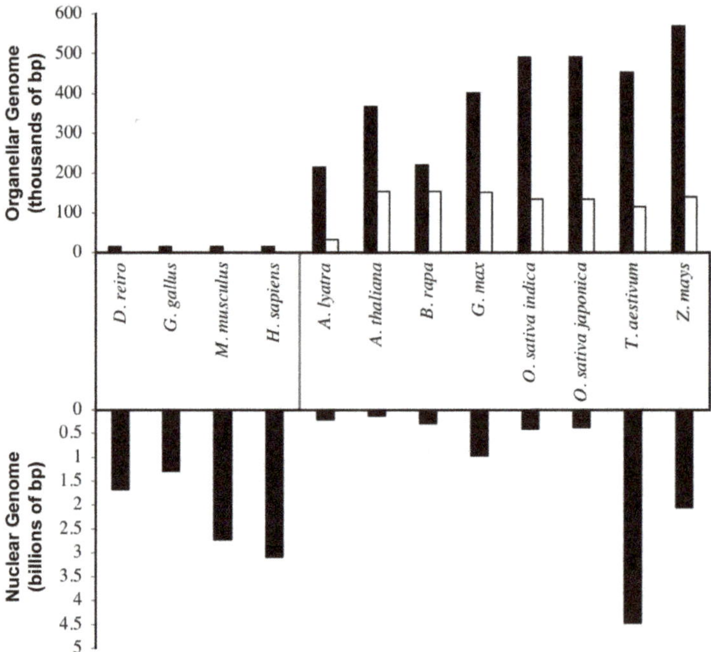

Figure 1. Nuclear and organelle genome sizes among different organisms. Mitochondrial genomes (dark bars in upper panel) among animals are compact and remarkably similar in size: ~16.5 kb. Plants, however, have mitochondrial genomes that dwarf those found in animals and vary in size from species to species. Chloroplast genomes vary less in size (open bars in upper panel) from organism to organism but still are relatively large compared to animal mitochondrial genomes. Organelle genome sizes do not correlate with nuclear genome size (bottom panel).

Compared to mitochondrial genomes, chloroplast genomes are fairly uniform, ranging between 120 and 160 kb in size, with some exceptions being as large as 2000 kb [17,18]. Higher plant chloroplast genomes possess large inverted repeats varying between 22–24 kb, accounting for ~16% of the chloroplast genome [19]. Surprisingly, removal of these repeats is associated with a higher frequency of recombination events and fewer nucleotide substitution events [20,21]. It is thought that these large inverted repeats in higher plants are involved in maintaining fidelity and correction of mutations or reducing errors in the cpDNA [21,22].

This uniformity suggests that genome reduction in chloroplasts might have taken place in a relatively short period of time soon after endosymbiosis [22]. Cytological observations indicate that cpDNA copy numbers seem to be related to plastid size, plastid type, developmental stage of the plant, and tissue type, and cpDNA copy number estimates per chloroplast range from several hundred to nearly two thousand per organelle [23,24]. It has been observed that the amount of cpDNA increases during the conversion of proplastids to mature chloroplasts (development), but it decreases during senescence (aging) [25–28]. This implies that reduction in cpDNA level can be tolerated to some extent and still maintain organelle functionality.

2.2. Genome Structure and Content

In both mitochondria and chloroplasts the DNA is associated with positively charged proteins in nucleoids [12]. Animal mtDNA consists of a singular circular molecule [29] and is very gene dense, with about 97% of the DNA coding for functional genes [30,31]. In most animals, the small non-coding region has important sequence elements for regulation of DNA replication and gene transcription [32]. One significant exception to this generalization can be seen in non-bilaterian animals, which possess large segments of non-coding DNA and varying levels of linear and circular DNA molecules [31].

For the most part, animal mitochondrial genomes encode the same 37 genes: two for rRNAs, 13 for proteins and 22 for tRNAs [10]. All 37 of these genes possess homologs in plants, fungi, and protists. To date, mtDNA gene content among animals only varies in nematodes [33], a bivalve [34], and cnidarians [35]. In these exceptions, there have been losses and gains of different mitochondrial genes, mostly tRNA genes.

Although plant mitochondrial genomes are mapped as circular molecules (master circles), circular molecules equal to a genome equivalent have only been observed in cultured liverwort cells [36]. Typically, plant mtDNA is observed primarily as large subgenomic linear molecules when observed by electron microscopy or pulsed field gel electrophoresis (PFGE). PFGE utilizes alternating direction of the current to allow separation of large DNA molecules or complex structures such as branched, lariat, rosette or catenated molecules, which are found in varying abundance in plant mtDNA preparations [37–39]. With PFGE, a large portion of the plant mtDNA remains trapped in highly complex arrangements near the wells [40]. Viewed by electron microscopy, these complex arrangements form DNA 'rosettes' and branched molecules, suggesting high levels of recombination. Other high molecular weight plant mtDNA simply does not enter the gel at all and has been theorized to be relaxed circle DNA, complex replication intermediates, or DNA bound to a matrix of other materials.

In contrast to animals, plant mtDNA contains many more genes and large portions of non-coding or undefined DNA [41]. A typical plant mitochondrial genome encodes anywhere between 50 and 100 genes [42]. The large genome size is at least partially due to the presence of non-coding DNA sequences, which consist of introns, repeats, and duplications of regions of the genome [41,43]. The known genes encode rRNA and tRNA genes as well as subunits for oxidative phosphorylation chain complexes [44]. The presence of these large non-coding DNA may have a role in lowering the mutation rate [45], as observed in *Arabidopsis thaliana* ecotypes Col-0 and C24. These two ecotypes have genetically identical mitochondrial genomes, but arrange their genes in different orders [46,47].

Chloroplast genomes exist primarily as homogeneous closed circle DNA molecules [48,49]. A small portion of these molecules also exist as circular dimers [19]. One exception to these observations can be seen among two species of brown algae [50]. In general, genes are conserved in most chloroplast genomes. For the most part, these consist of rRNA, tRNA, and genes involved in photosynthesis [17,18]. Loss of genes seems to be the only difference in gene content when comparing genomes. In these cases, essential genes have been lost from the chloroplast genome and transferred to the nucleus. Considering the possibility of multiple independent endosymbiotic events, it is interesting to observe the relatively conserved number and type of genes found in chloroplast genomes.

3. Organelle DNA Replication

3.1. Models for Organelle DNA Replication

3.1.1. Animal Mitochondria

Several modes of DNA replication in animals have been proposed (Figure 2). These include rolling circle, theta replication, strand-displacement, and RITOLS (Ribonucleotide Incorporation ThroughOut the Lagging Strand)/bootlace [51]. Rolling circle replication assures efficient reproduction of genomes exploiting a bacteriophage-like mechanism. Theta replication is the predominant replication mode among invertebrates, although nematodes have been observed to employ rolling circle DNA replication [52].

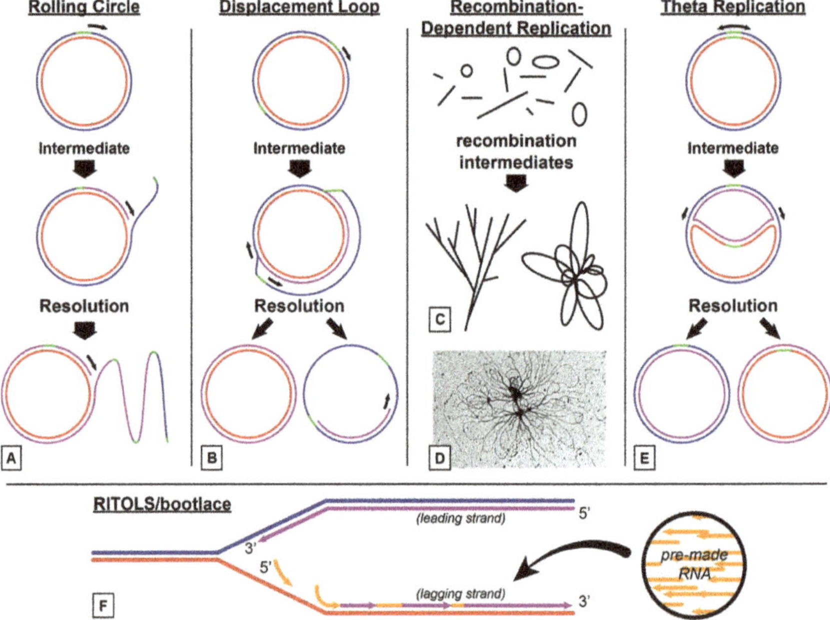

Figure 2. Proposed DNA replication mechanisms for mitochondrial DNA. (**A**) Rolling circle replication involves unidirectional replication after nicking of one DNA strand. DNA replication continues along the circular molecule displacing the nicked strand. Upon reaching the initial start site, the displaced strand may be nicked and ligated to form a new single stranded circular molecule or synthesis may continue, creating a linear concatemeric molecule which is later converted into multiple single stranded circular copies of the parent molecule. (**B**) Displacement loop (D-loop) replication proceeds unidirectionally by synthesis of an RNA primer that displaces one of the DNA strands. Upon synthesizing a certain portion of the genome (commonly 2/3) a second origin site is exposed as a single strand, which triggers DNA synthesis in the opposite direction. By the time the first double stranded

DNA molecule is finished, synthesis on the parent strand is still ongoing. Once replication reaches the initial start site, the parent strand is displaced as a single stranded circular DNA molecule. The single stranded circular molecules formed by rolling circle and displacement loop replication are later turned into double stranded copies by DNA replication machinery. (**C**) Recombination-dependent replication (RDR) involves the use of many linear and circular pieces of DNA that share homology. These pieces recombine to form branched linear and "rosette" like intermediates that are copied and replicated by DNA machinery. (**D**) Electron micrograph image of DNA forming a "rosette" that is likely the result of recombination. (**E**) Theta replication is so named because of the intermediate it forms as a result of bi-directional DNA replication. Replication initiates bi-directionally at an origin of replication, forming two replication forks. When these replication forks meet, the two double stranded circular molecules are separated. (**F**) The RITOLS (Ribonucleotide Incorporation ThroughOut the Lagging Strand)/bootlace strategy of replication involves the lagging strand of a replication fork. While the leading strand replicates normally, free pre-synthesized RNA molecules in the mitochondria (indicated by the arrow) hybridize to the lagging strand of the mtDNA starting from the 3′ end of the RNA and proceeding in the 5′ direction. Gaps are filled in and the primers are removed by the DNA replication machinery.

In vertebrates, two methods of mtDNA replication are currently accepted (Figure 2). The first is strand-displacement, or D-loop replication [53]. RITOLS/bootlace replication is a variation of D-loop replication. RITOLS was coined after scientists observed replication intermediates that were resistant to DNA endonucleases but sensitive to RNaseH [54].

Animals utilize a simple minimal DNA replisome for the replication of mtDNA. This replisome is made up of TWINKLE DNA helicase and DNA polymerase Polγ [55]. These two enzymes are somewhat processive and can synthesize molecules about 2 kb in length. The addition of single stranded binding protein (SSB) to TWINKLE and Polγ increases the processivity of this replisome to generate the genome sized molecules of 16 kb.

3.1.2. Plant Mitochondria

Plants most likely employ multiple mechanisms for replication of the mtDNA due to the complex structure of the mitochondrial genome. The structure of plant mtDNA makes strand displacement (D-loop) replication implausible, although there is one report of this mechanism observed in petunia flowers [56]. Rolling circle replication has also been observed in *Chenopodium album*, suggesting it could be a common replication mode in other plant species as well [57,58]. However, it is not possible to predict the exact mechanism for plant mtDNA replication, due to the large amount of non-coding DNA and the complex DNA structures observed, as mentioned above. Many scientists have proposed, based on the available information, that the main methods plants use for mtDNA replication include recombination-dependent replication (RDR) and recombination independent rolling circle replication. The mitochondrial resolvase Cce1 has been shown to play a role in mtDNA segregation [59].

3.1.3. Plant Chloroplasts

Replication of cpDNA is better understood than plant mitochondrial DNA replication. Chloroplasts utilize a double displacement loop strategy to initiate DNA replication [60]. The two displacement loops begin on opposite strands and begin replicating unidirectionally towards each other until they join to create a bidirectional replication bubble [61,62]. At this point, the displacement loops fuse, forming a Cairns or theta structure and DNA replication continues bidirectionally until two daughter molecules are created. Rolling circle and recombination-dependent replication have also been proposed for cpDNA [24,61,62]. MOC1 has been identified as a Holliday junction (recombination intermediate) resolvase that mediates chloroplast nucleoid segregation [63].

Some exceptions to the double D-loop replication model have been proposed. For example, *Chlamydomonas* and *Oenothera* possess two displacement loops, but discontinuous DNA replication begins shortly after initiation rather than after the fusion of the two D-loops [64,65]. *Euglena* possesses only one origin of replication site and appears to replicate bidirectionally from this site rather than forming two displacement loops [66].

3.2. Similarity to T7 Bacteriophage

Plant organelles likely mimic the minimal DNA replisome of T7 phage (Figure 3). The T7 DNA replisome consists of proteins gp5 (T7 DNA polymerase), gp4 (DNA helicase/primase) and gp2.5 (DNA single stranded binding protein). *E. coli* thioredoxin also binds to gp5 to increase the processivity of the enzyme [67]. Animal mitochondria use a similar system consisting of DNA Polγ, Twinkle, and SSB1 protein [55]. Since plant organelles possess the same proteins, one could logically assume that the same replisome is tasked with maintaining and replicating DNA in chloroplasts and plant mitochondria. However, while Twinkle knockouts in animals are lethal, Twinkle knockouts in plants lead to no distinguishable phenotype. Genome copy numbers in organelles also remain unchanged (S.A.M. and B.L.N., unpublished data). This contradicts the idea that, like with phage T7 and in animal mitochondria, a Twinkle-Pol1A/B replisome is the main driver of DNA synthesis in plant organelles. This also highlights the likelihood of plants utilizing multiple methods to replicate the organellar genomes rather than depending on a single mechanism.

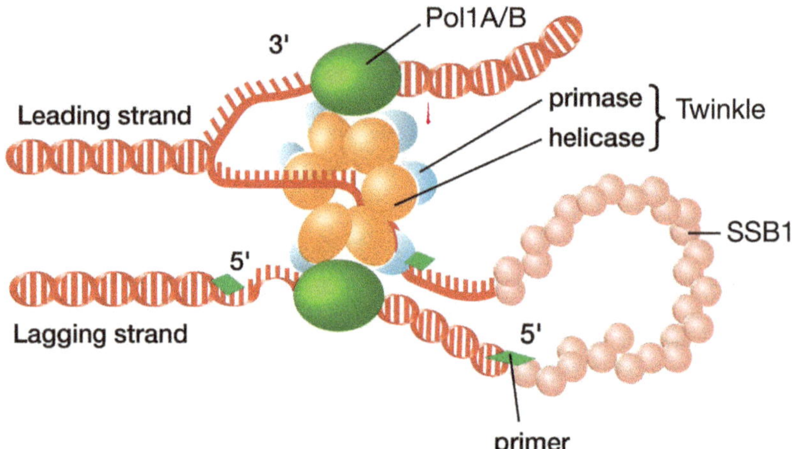

Figure 3. Theoretical model of the plant organellar DNA replisome. Four proteins are most likely involved in the minimal plant organellar DNA replisome, including Pol1A or Pol1B DNA polymerase, Twinkle DNA helicase/primase, and SSB1 single stranded binding protein. This model is similar to the replisome used by T7 phage which includes the proteins gp5 (DNA polymerase), gp4 (helicase/primase), and gp2.5 (single stranded binding protein). Adapted from Wikipedia file: phage T7 replication machinery.png, created 1 February 2015; https://creativecommons.org/licenses/by-sa/4.0/.

4. Proteins Involved in Plant Mitochondrial and Chloroplast DNA Replication

4.1. Organelle DNA Replication Proteins

The genomes in both mitochondria and chloroplasts are complexed with positively charged proteins in nucleoids [12,68], and this is the form of the DNA that is replicated in the organelles. Key functions required for DNA replication include polymerization, DNA unwinding, priming, strand separation, recombination, and ligation. These functions are carried out by nuclear encoded proteins that target to either the mitochondria, chloroplasts, or both. For the sake of simplicity, we will only discuss those replication proteins described in *Arabidopsis,* as their homologs exist in all vascular plants (Table 1). An interesting point to mention is that DNA replication proteins in plant organelles have different phylogenetic sources. For example, the DNA polymerases are bacterial in origin, while Twinkle helicase-primase and RNA polymerases are phage-like.

Table 1. Proteins involved in plant organellar DNA replication.

Function	Protein Name	TAIR	Homology	Localization*	Ref.
DNA polymerase	Pol1A or Pol gamma 2	At1g50840	Bacterial	M, P	[69–71]
	Pol1B or Pol gamma 2	At3g20540	Bacterial	M, P	[69–71]
Helicase	Twinkle	At1g30680	Phage	M, P	[72,73]
	DNA2	At1g08840	Mammalian	?	[74,75]
Priming	Twinkle	At1g30680	Phage	M, P	[72,73]
	RNA polymerase				
	RpoT1	At1g68990	Phage	M	
	RpoT2	At5g15700	Phage	M, P	
	RpoT3	At2g24120	Phage	P	[76–79]
	RpoA	AtCg00740	Bacterial	P	
	RpoB	AtCg00190	Bacterial	P	
	RpoC1	AtCg00180	Bacterial	P	
	RpoC2	AtCg00170	Bacterial	P	
	RNaseH				
	AtRNH1B	At5g51080			[80]
Primer Removal	AtRNH1C	At1g24090	Bacterial	M, P	[80]
	EXO1	?	Bacterial	P	[68]
	EXO2	?			[68]
ssDNA binding, recombination monitoring	SSB1	At4g11060	Bacterial	M, P	[44,81]
	SSB2	At3g18580	Bacterial	M	[44]
	OSB1	At3g18580	Bacterial-like, but unique to plants	M	[82]
	OSB2	At4g20010	Unique to plants	P	[44]
	OSB3	At5g44785	Unique to plants	M, P	[77,82]
	OSB4	At1g31010	Unique to plants	M	[82]
	WHY1	At1g14410	Unique to plants	P	[83–85]
	WHY2	At1g71260	Unique to plants	M	[83–85]
	WHY3	At2g02740	Unique to plants	P	[83–85]
	ODB1	At1g71310		M, N?	
	ODB2	At5g47870		P, N?	
Recombination	RecA1	At1g79050	Bacterial	P	[86]
	RecA2	At2g19490	Bacterial	M, P	[86]
	RecA3	At3g10140	Bacterial	M	[86,87]
	MSH1	At3g24320	Bacterial	M, P	[88]
Topoisomerase	Topoisomerase I	At4g31210	Bacterial	M, P	[77]
	DNA Gyrase A	At3g10690	Bacterial	M, P	[89]
	DNA Gyrase B1	At3g10270	Bacterial	P	[89]
	DNA Gyrase B2	At5g04130	Bacterial	M	[89]
	DNA Gyrase B3	At5g04110	Eukaryotic	N	[89]
Ligation	LIG1	At1g08130	Bacterial	M, N	[44]

* Localization to M (mitochondria), P (plastids), or N (nucleus). ? = unknown or potential localization.

4.2. DNA Polymerases

To date, two organellar DNA polymerases, Pol1A and Pol1B, resembling bacterial DNA Pol1, have been discovered in both mitochondria and chloroplasts. Although Pol1A and Pol1B are similar to each other, notable differences between the two have been observed. Pol1B knockout plants were shown to have fewer genome copy numbers per organelle and grew slowly [90], whereas the ΔPol1B mutant showed increased sensitivity to double stranded DNA breaks, suggesting a predominant role of PolB in cpDNA damage repair [91]. Recent studies show that Pol1A replicates DNA with high fidelity and has an increased ability to displace DNA when replicating over short single stranded gaps of DNA [92,93]. Efforts to create a double mutant for both DNA polymerases has not been successful yet, suggesting the essentiality of these polymerases for plant survival. However, heterozygous plants containing a single copy of either Pol1A or Pol1B were able to grow to maturity.

Both Pol1A and Pol1B have been shown to replicate an entire genome equivalent in mitochondria and chloroplasts with a greater efficiency than the microbial DNA Pol1 [91,94,95]. Interestingly, recombinant versions of *E. coli* DNA Pol1 were able to bind with thioredoxin, resulting in a dramatic increase in processivity [96]. It is quite possible that plant organelle DNA polymerases may also bind thioredoxin to achieve high processivity; however, this has yet to be shown experimentally.

Both Pol1A and Pol1B are able to bypass DNA lesions and continue replicating DNA [97,98]. These types of DNA polymerases do not replicate DNA with a high degree of fidelity as observed for human POLQ and yeast DNA Polymerases [97,98]. One possible explanation could be the presence of 3′-5′ proofreading exonuclease domains, which are absent in typical translesion synthesis DNA polymerases [98].

Three unique amino acid insertions (insertions 1, 2 and 3) have been identified in both Pol1A and Pol1B that confer translesion activity. Two of these insertions exist in a domain that stabilizes exiting DNA and increases processivity of the polymerase (insertions 1 and 2) and the third resides in a domain that properly positions the template strand (insertion 3). These appear to be flexible elements as mutants lacking all three of these insertions are still able to synthesize DNA [98]. The translesion activity of Pol1A and Pol1B is negatively affected by the removal of insertions 1 and 3, indicating that these enzymes acquired translesion activity through evolution and the acquisition of these insertions.

4.3. DNA Unwinding

In *Arabidopsis*, a Twinkle DNA helicase/primase has been studied to determine its function and properties [72,99]. *Arabidopsis* also employs several gyrases to relieve tension in the replicating DNA molecules. There may be other DNA unwinding enzymes active in the organelles, but these have not been identified. Twinkle (T7 gp4-like protein with intramitochondrial nucleoid + localization) is similar to the bacteriophage T7 gp4 DNA primase/helicase protein and gets its name from Spelbrink et al, who noted that the protein fused to EGFP produced punctate fluorescence patterns similar to twinkling stars [100]. In *Arabidopsis* and T7 phage, Twinkle possesses a zinc finger domain that allows the protein to bind DNA and synthesize RNA primers for replication. In humans, amino acid changes to this area of the protein have removed the zinc finger domain and its priming ability [101]. However, in plants, both of these capabilities remain intact [67,72,99]. When compared to T7 gp4, *Arabidopsis* Twinkle has a slight extension between the primase and helicase domains when compared to phage gp4 and a longer N-terminal region. Twinkle localizes to both chloroplasts and mitochondria in plants.

Arabidopsis also possesses a truncated form of Twinkle referred to as Twinky (At1g30660). This truncation lacks the C-terminal helicase domain of Twinkle but maintains the primase domain. No work on Twinky has been published, and whether it is active in priming DNA or is simply a pseudogene is not confirmed.

One possible alternative DNA helicase in plants is DNA2 [102]. JHS1 in *Arabidopsis* is homologous to human and yeast nuclease/helicase DNA2. In humans and yeast, DNA2 cleaves at the junction between single stranded DNA (ssDNA) and double stranded DNA (dsDNA) at the base of a DNA flap [102]. Experiments with human DNA2 and DNA POLγ have shown a positive interaction and the ability to unwind DNA without cleaving the D-loop structure observed in human mitochondria [75]. In humans, DNA2 localizes to the nucleus and to mitochondria [74]. *Arabidopsis* DNA2 has not yet been shown to localize to either organelle. DNA2 is essential in humans, yeast, and *Arabidopsis* as mutations lead to a lethal phenotype [74,75,103], while its role in plants and plant mtDNA replication has yet to be defined [104].

4.4. Organelle DNA Gyrases

Research studying organellar gyrases is almost nonexistent; however, there is confirmation of gyrase A (GYRA) in *Arabidopsis* localizing to both chloroplasts and mitochondria. *Arabidopsis* also has two other gyrases that localize either to chloroplasts (GYRB1) or mitochondria (GYRB2) [89]. Disruption of any of these gyrase genes leads to a lethal phenotype. Depletion of DNA gyrase in *Nicotiana benthamiana* result in abnormal nucleoids, chloroplasts, and mitochondria. The role of the organellar gyrases in DNA replication is presently unknown [105]

4.5. Priming of DNA Synthesis

Arabidopsis may utilize Twinkle and RNA polymerase (RNAP) to prime organellar DNA replication. As mentioned previously, Twinkle is similar to T7 gp4 protein and possesses both DNA helicase and primase activities. Using Twinkle to prime organellar DNA synthesis is unique to plants, as animals utilize RNA polymerase to prime their mtDNA [106,107]. Nonetheless, plants do contain organellar RNA polymerases that could complement the activity of Twinkle [78,79].

Twinkle uses a unique recognition sequence to begin ribonucleotide synthesis and appears to prefer cytosine and guanine incorporation over uracil and adenine [99]. The recognition sequence is 5′-(G/C)GGA-3′ where the underlined nucleotides are cryptic. If either of the cryptic nucleotides or the guanine directly upstream from them are substituted, RNA synthesis is abolished. This is unique from other DNA primases, in that two cryptic nucleotides are required for synthesis whereas other primases often require one. The exact mechanism of Twinkle association with template DNA is not fully understood.

Twinkle preferentially incorporates CTP and GTP, which is curious as nearly all plant mitochondrial and chloroplast genomes are highly A/T rich [108]. Why then would a plant organellar primase preferentially incorporate CTP and GTP? One theory points to *Aquifex aeolicus*, a primitive thermophilic bacteria, in which primer synthesis is initiated from a trinucleotide sequence composed of cytosines and guanines much like *Arabidopsis* Twinkle [109]. This G–C rich sequence is hypothesized to provide stability during primer extension. Similarly, plants may rely on the stability of the template sequence 5′-(G/C)GGA-3′ paired with preferential CTP and GTP incorporation to provide thermodynamic stability. Another leading theory is that preferential incorporation of CTP and GTP aid in determining Okazaki fragment length [99].

The co-evolution of nuclear, plastid, and mitochondrial genomes in plants has led to an interesting arrangement of RNA polymerases (RNAP) in the organelles. Unlike animal mitochondria, which utilize a single RNA polymerase [110], plant organelles require multiple RNAPs: at least two for plastids and one for mitochondria. These genes are designated as "*RpoT*" genes, the "*T*" indicating their similarity to the single subunit RNA polymerases of T3 and T7 phage [111]. RNAP that targets to mitochondria are designated RpoTm and those that target to plastids are called RpoTp. RpoTmp represent RNAP that target to both organelles. Different species may possess multiple copies of these nuclear encoded organellar proteins, but the earliest phylogenetic versions of these enzymes exist in the waterlily *Nuphar advena*, a basal angiosperm [112]. All RNAPs in mitochondria are nuclear encoded, while plastids use both nuclear and plastid-encoded RNAPs. Extensive research has been performed on how plant RNA polymerases recognize promoters and transcribe genes. This portion of the review focuses on the potential RNAP has to prime DNA for synthesis rather than its role in transcribing genes.

Three single-subunit mitochondrial RpoT genes have been identified in *Arabidopsis*; however, only two have been proven to localize to mitochondria [79,111]. A duplication of one of these genes has led to the creation of a RpoTmp. How these enzymes coordinate synthesis of RNA is largely unexplored, although some research suggests RpoTmp is responsible for gene synthesis in early seedling development and RpoTm and RpoTp take over once the plant has fully developed [113].

In *Arabidopsis*, plastids require at least two RNAPs: nuclear encoded RNAP (NEP), and plastid encoded bacterial-like RNAP (PEP). The NEP and PEP versions of RNAP are distinct and do not share subunits [114]. The NEP is homologous to the phage-like single subunit RNA polymerases found in mitochondria. This RNAP is represented as RpoTp and is thought to be a duplication of the mitochondrial RpoTm. NEP isolated from *P. sativum* seems to act more like a primase than an RNAP [115], as it makes primers that are too short for transcription but much larger than other RNAPs. Like other RNAPs, it also showed preferential binding to single-stranded rather than double stranded DNA. The enzyme was also found to be resistant to inhibition by tagetitoxin, a specific inhibitor of chloroplast-encoded RNA polymerase, as well as polyclonal antibodies specific to purified pea chloroplast RNAP. These findings suggest that plastids and probably mitochondria possess an RNAP gene that functions as a DNA primase, although further research on this topic is needed.

Unlike the mitochondrial phage-like RNA polymerases, the PEP is made up of multiple subunits that share homology with the core subunits of *E. coli* RNAP: α, β, β′, and β″. These subunits are encoded from the genes *rpoA*, *rpoB*, *rpoC1*, and *rpoC2*, respectively. In addition, several nuclear-encoded sigma factors form the PEP holoenzyme and provide promoter recognition for the plastid encoded subunits [76]. In agreement with the theory of endosymbiosis, the core enzyme of the NEP is also homologous to multi subunit RNA polymerases of cyanobacteria [111].

In humans and yeast, RNAPs are required to initiate and prime DNA for replication in mitochondria [106,107]. This makes sense when observing that Twinkle, a helicase-primase, is present in animals but lacks the primase activity observed in both phage and plants. Therefore, plants may also use their organellar RNA polymerases to prime DNA for synthesis. Unfortunately, the ability and scale on which this actually happens is grossly understudied, most likely due to the assumption that organellar DNA is primed by mimicking the simple replisome found in T7 phage. However, unlike animals in which mutation of Twinkle helicase/primase is embryo-lethal, plants with Twinkle knock-out mutations grow well, and display no phenotypic defects [116]. Therefore, the ability of RNA polymerase to prime DNA for synthesis may be extremely important to plants and could be a fruitful area of research.

4.6. Primer Removal

In *E. coli*, RNA primers are removed from DNA–RNA hybrids by the 5′-3′ exonuclease activity of DNA polymerase I. Since Pol1A and Pol1B lack 5′-3′ exonuclease activity, primer removal must be carried out by another enzyme. Recent work has shown RNase H-like activity both in mitochondria and chloroplasts [80]. In addition to RNase H, two exonucleases with homology to the 5′-3′ exonuclease domain of *E. coli* DNA polymerase I (5′-3′ EXO1 and 2) have also been predicted to localize to chloroplasts or mitochondria [68].

4.7. Single-Stranded DNA Binding Proteins

Plants utilize at least two types of single stranded DNA binding proteins (SSBs) in their organelles. The first one is similar to bacterial SSBs. *Arabidopsis* encodes at least two of these genes, called SSB1 and SSB2, although little is known about SSB2 [44,81]. SSB1 functions at replication forks by coating single stranded DNA to prevent fork collapse. This protein has been shown to localize to both mitochondria and chloroplasts, and stimulates bacterial RecA activity [39,44,81]. RecA is a bacterial protein involved in homologous recombination and strand invasion, and is discussed in greater detail below.

The second class of SSB proteins is called organellar single-stranded DNA binding proteins (OSB). OSB proteins are distinct from the bacterial SSB versions and are unique to plant organelles. At least four OSB genes are transcribed in *Arabidopsis* with OSB proteins localizing to both the mitochondria and chloroplasts. Although the function of these molecules has not been completely detailed, mutants for OSB1 accumulate mtDNA homologous recombination products. In subsequent generations, these products segregate into separate plant lines where one of the homologous recombination products is predominant. If OSB1 activity is restored the plants segregate into separate line they will revert to wild type configurations of mtDNA. However, if segregation has already occurred, restoration of OSB1 activity does not restore plants to wild type mtDNA configurations [82]. Therefore, OSB proteins are most likely involved in recombination surveillance and preventing transmission of incorrect recombination products to newly formed mitochondria.

In addition to OSB proteins, plants also utilize Whirly (WHY) and organellar DNA binding (ODB) proteins. WHY proteins form tetramers that take on the appearance of a whirligig, hence the name Whirly. There are three Whirly proteins—WHY1, WHY2, and WHY3 [83]. WHY1 and WHY3 localize to chloroplasts while WHY2 localizes to mitochondria. Whirly appears to have expanded roles from OSB proteins. Like OSB, WHY proteins also appear to be involved in recombination surveillance and have also been shown to associate with RNA molecules. Some WHY proteins have been associated with double stranded DNA repair [84,85] and regulation of defense genes in response to pathogens [117].

4.8. DNA Recombination

As mentioned in the previous section, OSB, ODB, and WHY proteins have been shown to participate in DNA repair in mitochondria and chloroplasts. In addition, plant organelles also contain proteins involved in homologous DNA recombination. There are two classes of proteins dedicated to recombination in plant organelles. One is a MutS homolog called MSH1, and the others are RecA homologs.

MutS from *E. coli* is part of the mismatch repair (MMR) pathway and corrects point mutations and small insertions and deletions by preventing recombination between partially homologous DNA sequences. MSH1 is a nuclear encoded gene and mutants display patchy green/white/yellow leaves symptomatic of dysfunctional chloroplasts, with a non-Mendelian inheritance pattern [118]. Due to this activity, this gene was originally called *chm* for *chloroplast mutator*. Later, it was discovered that *chm* mutants cause rearrangements to the mitochondrial genome that lead to the observed phenotypes and defective chloroplasts. Despite extensive searching, no mutation or rearrangement of the plastid genome has been observed in *chm* mutants [119]. Being homologous to MutS from *E. coli*, the gene was consequently renamed MSH1 and mutants designated as *msh1* plants [120].

Insertion mutations of yeast *msh1* lead to a petite phenotype indicative of mitochondrial dysfunction. Mutation of yeast *msh1* is also accompanied by large-scale point mutations and rearrangements in the mitochondrial genome [121]. Interestingly, plant MSH1 mutants do not accumulate point mutations over time, suggesting that the plant MSH1 specializes primarily in recombination and is not essential for correcting mismatches.

E. coli utilizes the adaptor protein MutL and endonuclease MutH to assist in the mismatch repair pathway, but no homologs of these proteins have been identified yet in plant organelles. Instead, plant MSH1 possesses three recognizable domains and three unknown domains to facilitate mismatch repair. These include a conserved DNA binding and mismatch recognition domain and an ATPase domain homologous to those in *E. coli* MutS. Point mutations to the ATPase and endonuclease domains of plant MSH1 led to the defective chloroplast phenotype [122], suggesting that plant MSH1 may provide mismatch recognition and base excision without the need for MutL or MutH homologs, although this has not been experimentally shown. Recent studies have shown that MSH1 suppresses homeologous recombination [123,124]. Plant MSH1 also has a unique GIY-YIG endonuclease domain, which binds specifically to the D-loop structure, suggesting that this protein may recognize and resolve mismatch-containing intermediates [125].

RecA facilitates homologous recombination by correctly pairing homologous sequences and promoting strand invasion. Eukaryote versions of this protein are called RAD51. All homologous recombination begins with strand invasion mediated by RecA family proteins, making this protein crucial for this type of repair. RecA functions by coating single stranded DNA at lesions to form presynaptic filaments. This complex will then search for homology within intact double stranded DNA. Once homology has been established, the presynaptic complex will destabilize the double stranded DNA promoting strand exchange and D-loop formation. Three RecA proteins have been identified in *Arabidopsis*: RecA1, RecA2, and RecA3. RecA1 localizes to plastids, RecA3 to mitochondria, and RecA2 to both organelles [86]. Mutations to RecA3 cause mitochondrial rearrangements distinct from those observed in MSH1 mutants. The rearrangements observed in RecA3 mutants are due to homologous recombination of repeated sequences in the mitochondrial genome. Reintroducing RecA3 into these mutants results in a reversal of this effect in most of the progeny by abolishing aberrant mitochondrial DNA molecules. RecA1 and RecA2 appear to be even more essential to homologous recombination as mutations in these genes cause a seed-lethal phenotype. This may be explained by the lack of the C-terminal domain found in both RecA1 and RecA2 as well as bacterial homologs. In bacteria, deletion of this C-terminal domain enhances the activity of RecA, suggesting its involvement in autoregulation.

4.9. DNA Ligation

The amount of information surrounding plant organellar DNA ligases is extremely limited. Unlike mitochondria, no DNA ligase has been confirmed or observed functioning in plastids, representing a potential avenue of research, as the activity of this enzyme in both organelles must be present. Four DNA ligase genes have been identified in the genome of *Arabidopsis*; however, only DNA ligase 1 (LIG1) has been identified in mitochondria [126]. Plant LIG1 knockouts are seedling-lethal and knockdown mutants display severe growth defects due to effects on the nuclear genome rather than the mitochondrial genome [127].

5. Conclusions

Genomes found in plant mitochondria and chloroplasts are essential for organelle function, but there is still relatively little known about how these genomes are replicated and maintained. This is especially true for plant mitochondria, which have a very large variation in genome size depending on the species, and there is considerable evidence that the genome exists as linear subgenomic molecules, raising questions as to how the integrity of the genetic information is maintained. Thus, plant mitochondrial genomes and their replication are much more complex than their animal counterparts. It is clear that for at least some replication functions more than one gene is present in *Arabidopsis*, suggesting the possibility of functional redundancy. For chloroplasts, although a DNA replication mechanism has been established, it is quite possible that more than one mechanism is involved, perhaps for different stages of growth or in response to different signals. Further research is needed to better understand the basic structure of the organelle genomes and how these DNA molecules are replicated. In addition, the mechanism(s) for maintaining genome copy number and regulation of replication initiation are not known and should be studied.

Author Contributions: Conceptualization of the ideas for this review were by S.A.M. and B.L.N. Writing, original draft preparation including figures and tables were by S.A.M. S.A.M., N.A. and B.L.N. were involved in writing—review and editing, and all coauthors have read and approved of the manuscript. Funding acquisition, B.L.N.

Funding: Research on plant organelle DNA replication and recombination in the Nielsen lab has been provided by a BYU Mentoring Environment Grant and the BYU Dept. of Microbiology & Molecular Biology. S.A.M. was funded by a BYU Graduate Studies Research Fellowship to study plant mitochondrial DNA genome maintenance. N.A. is thankful to HEC, Pakistan, for providing funds in his lab.

Acknowledgments: We acknowledge the assistance of several undergraduate students on various projects in the BLN research laboratory related to plant mitochondrial DNA replication and recombination.

Conflicts of Interest: The authors declare no conflict of interest. The funders had no role in the design of the study; in the collection, analyses, or interpretation of data; in the writing of the manuscript, or in the decision to publish the results.

References

1. Hooke, R. *Micrographia: Or Some Physiological Descriptions of Minute Bodies Made by Magnifying Glasses with Observations and Inquiries Thereupon*; Facsimile Edition; Science Heritage, Ltd.: Lincolnwood, IL, USA, 1987; 323p.
2. Harris, H. *The Birth of the Cell*; Yale University Press: New Haven, CT, USA, 1999; p. 212.
3. Ernster, L.; Schatz, G. Mitochondria—A Historical Review. *J. Cell Biol.* **1981**, *91*, S227–S255. [CrossRef] [PubMed]
4. Schimper, A.F.W. Ueber die Entwickelung der Chlorophyllkörner und Farbkörper. *Botanische Zeitung* **1883**, *41*, 105–160.
5. Timmis, J.N.; Ayliffe, M.A.; Huang, C.Y.; Martin, W. Endosymbiotic gene transfer: Organelle genomes forge eukaryotic chromosomes. *Nat. Rev. Genet.* **2004**, *5*, 123–135. [CrossRef] [PubMed]
6. Kutschera, U.; Niklas, K.J. Endosymbiosis, cell evolution, and speciation. *Theory Biosci.* **2005**, *124*, 1–24. [CrossRef] [PubMed]
7. Zimorski, V.; Ku, C.; Martin, W.F.; Gould, S.B. Endosymbiotic theory for organelle origins. *Curr. Opin. Microbiol.* **2014**, *22*, 38–48. [CrossRef] [PubMed]
8. Bergthorsson, U.; Adams, K.L.; Thomason, B.; Palmer, J.D. Widespread horizontal transfer of mitochondrial genes in flowering plants. *Nature* **2003**, *424*, 197–201. [CrossRef]
9. Blanchard, J.L.; Schmidt, G.W. Pervasive migration of organellar DNA to the nucleus in plants. *J. Mol. Evol.* **1995**, *41*, 397–406. [CrossRef]
10. Boore, J.L. Animal mitochondrial genomes. *Nucleic Acids Res.* **1999**, *27*, 1767–1780. [CrossRef]
11. Iborra, F.J.; Kimura, H.; Cook, P.R. The functional organization of mitochondrial genomes in human cells. *BMC Biol.* **2004**, *2*, 9. [CrossRef]
12. Kukat, C.; Wurm, C.A.; Spahr, H.; Falkenberg, M.; Larsson, N.G.; Jakobs, S. Super-resolution microscopy reveals that mammalian mitochondrial nucleoids have a uniform size and frequently contain a single copy of mtDNA. *Proc. Natl. Acad. Sci. USA* **2011**, *108*, 13534–13539. [CrossRef]
13. Montier, L.L.C.; Deng, J.J.; Bai, Y.D. Number matters: Control of mammalian mitochondrial DNA copy number. *J. Genet. Genom.* **2009**, *36*, 125–131. [CrossRef]
14. Fauron, C.; Allen, J.; Clifton, S.; Newton, K. Plant Mitochondrial Genomes. In *Molecular Biology and Biotechnology of Plant Organelles: Chloroplasts and Mitochondria*; Daniell, H., Chase, C., Eds.; Springer: Dordrecht, The Netherlands, 2004; pp. 151–177. [CrossRef]
15. Oldenburg, D.J.; Kumar, R.A.; Bendich, A.J. The amount and integrity of mt DNA in maize decline with development. *Planta* **2013**, *237*, 603–617. [CrossRef]
16. Preuten, T.; Cincu, E.; Fuchs, J.; Zoschke, R.; Liere, K.; Borner, T. Fewer genes than organelles: Extremely low and variable gene copy numbers in mitochondria of somatic plant cells. *Plant J.* **2010**, *64*, 948–959. [CrossRef]
17. Daniell, H.; Lin, C.S.; Yu, M.; Chang, W.J. Chloroplast genomes: Diversity, evolution, and applications in genetic engineering. *Genome Biol.* **2016**, *17*. [CrossRef]
18. Palmer, J.D. Comparative Organization of Chloroplast Genomes. *Annu. Rev. Genet.* **1985**, *19*, 325–354. [CrossRef]
19. Kolodner, R.; Tewari, K.K. Inverted Repeats in Chloroplast DNA from Higher-Plants. *Proc. Natl. Acad. Sci. USA* **1979**, *76*, 41–45. [CrossRef]
20. Palmer, J.D.; Thompson, W.F. Chloroplast DNA rearrangements are more frequent when a large inverted repeat sequence is lost. *Cell* **1982**, *29*, 537–550. [CrossRef]
21. Zhu, A.; Guo, W.; Gupta, S.; Fan, W.; Mower, J.P. Evolutionary dynamics of the plastid inverted repeat: The effects of expansion, contraction, and loss on substitution rates. *New Phytol.* **2016**, *209*, 1747–1756. [CrossRef]
22. Gray, M.W. The Bacterial Ancestry of Plastids and Mitochondria. *BioScience* **1983**, *33*, 693–699. [CrossRef]
23. Zoschke, R.; Liere, K.; Borner, T. From seedling to mature plant: Arabidopsis plastidial genome copy number, RNA accumulation and transcription are differentially regulated during leaf development. *Plant J.* **2007**, *50*, 710–722. [CrossRef]
24. Zheng, Q.; Oldenburg, D.J.; Bendich, A.J. Independent effects of leaf growth and light on the development of the plastid and its DNA content in Zea species. *J. Exp. Bot.* **2011**, *62*, 2715–2730. [CrossRef]
25. Udy, D.B.; Belcher, S.; Williams-Carrier, R.; Gualberto, J.M.; Barkan, A. Effects of Reduced Chloroplast Gene Copy Number on Chloroplast Gene Expression in Maize. *Plant Physiol.* **2012**, *160*, 1420–1431. [CrossRef]

26. Shaver, J.M.; Oldenburg, D.J.; Bendich, A.J. Changes in chloroplast DNA during development in tobacco, Medicago truncatula, pea, and maize. *Planta* **2006**, *224*, 72–82. [CrossRef]

27. Rowan, B.A.; Oldenburg, D.J.; Bendich, A.J. The demise of chloroplast DNA in Arabidopsis. *Curr. Genet.* **2004**, *46*, 176–181. [CrossRef]

28. Rowan, B.A.; Oldenburg, D.J.; Bendich, A.J. A multiple-method approach reveals a declining amount of chloroplast DNA during development in Arabidopsis. *BMC Plant Biol.* **2009**, *9*. [CrossRef]

29. Clayton, D.A. Replication of animal mitochondrial DNA. *Cell* **1982**, *28*, 693–705. [CrossRef]

30. Anderson, S.; Bankier, A.T.; Barrell, B.G.; Debruijn, M.H.L.; Coulson, A.R.; Drouin, J.; Eperon, I.C.; Nierlich, D.P.; Roe, B.A.; Sanger, F.; et al. Sequence and Organization of the Human Mitochondrial Genome. *Nature* **1981**, *290*, 457–465. [CrossRef]

31. Lavrov, D.V.; Pett, W. Animal Mitochondrial DNA as We Do Not Know It: Mt-Genome Organization and Evolution in Nonbilaterian Lineages. *Genome Biol. Evol.* **2016**, *8*, 2896–2913. [CrossRef]

32. Shadel, G.S.; Clayton, D.A. Mitochondrial DNA maintenance in vertebrates. *Annu. Rev. Biochem.* **1997**, *66*, 409–435. [CrossRef]

33. Okimoto, R.; Macfarlane, J.L.; Clary, D.O.; Wolstenholme, D.R. The mitochondrial genomes of two nematodes, Caenorhabditis elegans and Ascaris suum. *Genetics* **1992**, *130*, 471–498.

34. Hoffmann, R.J.; Boore, J.L.; Brown, W.M. A Novel Mitochondrial Genome Organization for the Blue Mussel, Mytilus-Edulis. *Genetics* **1992**, *131*, 397–412. [PubMed]

35. Beagley, C.T.; Macfarlane, J.L.; PontKingdon, G.A.; Okimoto, R.; Okada, N.A.; Wolstenholme, D.R. Mitochondrial genomes of anthozoa (Cnidaria). *Prog. Cell Res.* **1995**, *5*, 149–153.

36. Oda, K.; Kohchi, T.; Ohyama, K. Mitochondrial-DNA of Marchantia-Polymorpha as a Single Circular Form with No Incorporation of Foreign DNA. *Biosci. Biotechnol. Biochem.* **1992**, *56*, 132–135. [CrossRef] [PubMed]

37. Backert, S.; Nielsen, B.L.; Borner, T. The mystery of the rings: Structure and replication of mitochondrial genomes from higher plants. *Trends Plant Sci.* **1997**, *2*, 477–483. [CrossRef]

38. Sloan, D.B. One ring to rule them all? Genome sequencing provides new insights into the 'master circle' model of plant mitochondrial DNA structure. *New Phytol.* **2013**, *200*, 978–985. [CrossRef] [PubMed]

39. Cupp, J.D.; Nielsen, B.L. Minireview: DNA replication in plant mitochondria. *Mitochondrion* **2014**, *19*, 231–237. [CrossRef] [PubMed]

40. Schuster, W.; Brennicke, A. The Plant Mitochondrial Genome—Physical Structure, Information-Content, Rna Editing, and Gene Migration to the Nucleus. *Annu. Rev. Plant Phys.* **1994**, *45*, 61–78. [CrossRef]

41. Palmer, J.D.; Adams, K.L.; Cho, Y.R.; Parkinson, C.L.; Qiu, Y.L.; Song, K.M. Dynamic evolution of plant mitochondrial genomes: Mobile genes and introns and highly variable mutation rates. *Proc. Natl. Acad. Sci. USA* **2000**, *97*, 6960–6966. [CrossRef] [PubMed]

42. Morley, S.A.; Nielsen, B.L. Plant mitochondrial DNA. *Front. Biosci-Landmrk* **2017**, *22*, 1023–1032. [CrossRef]

43. Burger, G.; Gray, M.W.; Lang, B.F. Mitochondrial genomes: Anything goes. *Trends Genet.* **2003**, *19*, 709–716. [CrossRef]

44. Gualberto, J.M.; Mileshina, D.; Wallet, C.; Niazi, A.K.; Weber-Lotfi, F.; Dietrich, A. The plant mitochondrial genome: Dynamics and maintenance. *Biochimie* **2014**, *100*, 107–120. [CrossRef] [PubMed]

45. Palmer, J.D.; Herbon, L.A. Plant Mitochondrial-DNA Evolves Rapidly in Structure, but Slowly in Sequence. *J. Mol. Evol.* **1988**, *28*, 87–97. [CrossRef] [PubMed]

46. Christensen, A.C. Plant Mitochondrial Genome Evolution Can Be Explained by DNA Repair Mechanisms. *Genome Biol. Evol.* **2013**, *5*, 1079–1086. [CrossRef]

47. Parsons, T.J.; Muniec, D.S.; Sullivan, K.; Woodyatt, N.; AllistonGreiner, R.; Wilson, M.R.; Berry, D.L.; Holland, K.A.; Weedn, V.W.; Gill, P.; et al. A high observed substitution rate in the human mitochondrial DNA control region. *Nat. Genet.* **1997**, *15*, 363–368. [CrossRef] [PubMed]

48. Sugiura, M. The chloroplast genome. *Essays Biochem.* **1995**, *30*, 49–57.

49. Palmer, J.D.; Stein, D.B. Conservation of Chloroplast Genome Structure among Vascular Plants. *Curr. Genet.* **1986**, *10*, 823–833. [CrossRef]

50. Dalmon, J.; Loiseaux, S.; Bazetoux, S. Heterogeneity of plastid DNA of two species of brown algae. *Plant Sci. Lett.* **1983**, *29*, 243–253. [CrossRef]

51. Ciesielski, G.L.; Oliveira, M.T.; Kaguni, L.S. Animal Mitochondrial DNA Replication. *Enzymes* **2016**, *39*, 255–292. [CrossRef]

52. Lewis, S.C.; Joers, P.; Willcox, S.; Griffith, J.D.; Jacobs, H.T.; Hyman, B.C. A Rolling Circle Replication Mechanism Produces Multimeric Lariats of Mitochondrial DNA in *Caenorhabditis Elegans*. *PLoS Genet.* **2015**, *11*. [CrossRef]

53. Fish, J.; Raule, N.; Attardi, G. Discovery of a major D-loop replication origin reveals two modes of human mtDNA synthesis. *Science* **2004**, *306*, 2098–2101. [CrossRef]

54. Holt, I.J.; Lorimer, H.E.; Jacobs, H.T. Coupled leading- and lagging-strand synthesis of mammalian mitochondrial DNA. *Cell* **2000**, *100*, 515–524. [CrossRef]

55. Korhonen, J.A.; Pham, X.H.; Pellegrini, M.; Falkenberg, M. Reconstitution of a minimal mtDNA replisome in vitro. *EMBO J.* **2004**, *23*, 2423–2429. [CrossRef]

56. Dehaas, J.M.; Hille, J.; Kors, F.; Vandermeer, B.; Kool, A.J.; Folkerts, O.; Nijkamp, H.J.J. 2 Potential Petunia-Hybrida Mitochondrial-DNA Replication Origins Show Structural and Invitro Functional Homology with the Animal Mitochondrial-DNA Heavy and Light Strand Replication Origins. *Curr. Genet.* **1991**, *20*, 503–513. [CrossRef]

57. Backert, S.; Dorfel, P.; Lurz, R.; Borner, T. Rolling-circle replication of mitochondrial DNA in the higher plant Chenopodium album (L). *Mol. Cell Biol.* **1996**, *16*, 6285–6294. [CrossRef]

58. Backert, S.; Borner, T. Phage T4-like intermediates of DNA replication and recombination in the mitochondria of the higher plant *Chenopodium album* (L.). *Curr. Genet.* **2000**, *37*, 304–314. [CrossRef]

59. Lockshon, D.; Zweifel, S.G.; Freeman-Cook, L.L.; Lorimer, H.E.; Brewer, B.J.; Fangman, W.L. A role for recombination junctions in the segregation of mitochondrial DNA in yeast. *Cell* **1995**, *81*, 947–955. [CrossRef]

60. Heinhorst, S.; Cannon, G.C. DNA-Replication in Chloroplasts. *J. Cell Sci.* **1993**, *104*, 1–9.

61. Kolodner, R.; Tewari, K.K. Presence of displacement loops in the covalently closed circular chloroplast deoxyribonucleic acid from higher plants. *J. Biol. Chem.* **1975**, *250*, 8840–8847.

62. Kunnimalaiyaan, M.; Nielsen, B.L. Chloroplast DNA replication: Mechanism, enzymes and replication origins. *J. Plant Biochem. Biot.* **1997**, *6*, 1–7. [CrossRef]

63. Kobayashi, Y.; Misumi, O.; Odahara, M.; Ishibashi, K.; Hirono, M.; Hidaka, K.; Endo, M.; Sugiyama, H.; Iwasaki, H.; Kuroiwa, T.; et al. Holliday junction resolvases mediate chloroplast nucleoid segregation. *Science* **2017**, *356*, 631–634. [CrossRef]

64. Chiu, W.L.; Sears, B.B. Electron-Microscopic Localization of Replication Origins in Oenothera Chloroplast DNA. *Mol. Gen. Genet.* **1992**, *232*, 33–39. [CrossRef]

65. Waddell, J.; Wang, X.M.; Wu, M. Electron-Microscopic Localization of the Chloroplast DNA Replicative Origins in Chlamydomonas-Reinhardii. *Nucleic Acids Res.* **1984**, *12*, 3843–3856. [CrossRef]

66. Ravelchapuis, P.; Heizmann, P.; Nigon, V. Electron-Microscopic Localization of the Replication Origin of Euglena-Gracilis Chloroplast DNA. *Nature* **1982**, *300*, 78–81. [CrossRef]

67. Lee, S.J.; Richardson, C.C. Choreography of bacteriophage T7 DNA replication. *Curr. Opin. Chem. Biol.* **2011**, *15*, 580–586. [CrossRef]

68. Sato, N.; Terasawa, K.; Miyajima, K.; Kabeya, Y. Organization, developmental dynamics, and evolution of plastid nucleoids. *Int. Rev. Cytol.* **2003**, *232*, 217–262. [CrossRef]

69. Elo, A.; Lyznik, A.; Gonzalez, D.O.; Kachman, S.D.; Mackenzie, S.A. Nuclear genes that encode mitochondrial proteins for DNA and RNA metabolism are clustered in the Arabidopsis genome. *Plant Cell* **2003**, *15*, 1619–1631. [CrossRef]

70. Ono, Y.; Sakai, A.; Takechi, K.; Takio, S.; Takusagawa, M.; Takano, H. NtPolI-like1 and NtPolI-like2, bacterial DNA polymerase I homologs isolated from BY-2 cultured tobacco cells, encode DNA polymerases engaged in DNA replication in both plastids and mitochondria. *Plant Cell Physiol.* **2007**, *48*, 1679–1692. [CrossRef]

71. Carrie, C.; Kuhn, K.; Murcha, M.W.; Duncan, O.; Small, I.D.; O'Toole, N.; Whelan, J. Approaches to defining dual-targeted proteins in Arabidopsis. *Plant J.* **2009**, *57*, 1128–1139. [CrossRef]

72. Diray-Arce, J.; Liu, B.; Cupp, J.D.; Hunt, T.; Nielsen, B.L. The Arabidopsis At1g30680 gene encodes a homologue to the phage T7 gp4 protein that has both DNA primase and DNA helicase activities. *BMC Plant Biol.* **2013**, *13*. [CrossRef]

73. Shutt, T.E.; Gray, M.W. Twinkle, the mitochondrial replicative DNA helicase, is widespread in the eukaryotic radiation and may also be the mitochondrial DNA primase in most eukaryotes. *J. Mol. Evol.* **2006**, *62*, 588–599. [CrossRef]

74. Duxin, J.P.; Dao, B.; Martinsson, P.; Rajala, N.; Guittat, L.; Campbell, J.L.; Spelbrink, J.N.; Stewart, S.A. Human Dna2 Is a nuclear and mitochondrial DNA maintenance protein. *Mol. Cell Biol.* **2009**, *29*, 4274–4282. [CrossRef]

75. Zheng, L.; Zhou, M.A.; Guo, Z.G.; Lu, H.M.; Qian, L.M.; Dai, H.F.; Qiu, J.Z.; Yakubovskaya, E.; Bogenhagen, D.F.; Demple, B.; et al. Human DNA2 Is a mitochondrial nuclease/helicase for efficient processing of DNA replication and repair intermediates. *Mol. Cell* **2008**, *32*, 325–336. [CrossRef]

76. Liere, K.; Weihe, A.; Borner, T. The transcription machineries of plant mitochondria and chloroplasts: Composition, function, and regulation. *J. Plant Physiol.* **2011**, *168*, 1345–1360. [CrossRef]

77. Carrie, C.; Small, I. A reevaluation of dual-targeting of proteins to mitochondria and chloroplasts. *Biochim. Biophys. Acta* **2013**, *1833*, 253–259. [CrossRef]

78. Hedtke, B.; Borner, T.; Weihe, A. One RNA polymerase serving two genomes. *EMBO Rep.* **2000**, *1*, 435–440. [CrossRef]

79. Hedtke, B.; Borner, T.; Weihe, A. Mitochondrial and chloroplast phage-type RNA polymerases in Arabidopsis. *Science* **1997**, *277*, 809–811. [CrossRef]

80. Yang, Z.; Hou, Q.C.; Cheng, L.L.; Xu, W.; Hong, Y.T.; Li, S.; Sun, Q.W. RNase H1 Cooperates with DNA Gyrases to Restrict R-Loops and Maintain Genome Integrity in Arabidopsis Chloroplasts. *Plant Cell* **2017**, *29*, 2478–2497. [CrossRef]

81. Edmondson, A.C.; Song, D.Q.; Alvarez, L.A.; Wall, M.K.; Almond, D.; McClellan, D.A.; Maxwell, A.; Nielsen, B.L. Characterization of a mitochondrially targeted single-stranded DNA-binding protein in Arabidopsis thaliana. *Mol. Genet. Genom.* **2005**, *273*, 115–122. [CrossRef]

82. Zaegel, V.; Guermann, B.; Le Ret, M.; Andres, C.; Meyer, D.; Erhardt, M.; Canaday, J.; Gualberto, J.M.; Imbault, P. The plant-specific ssDNA binding protein OSB1 is involved in the stoichiometric transmission of mitochondrial DNA in Arabidopsis. *Plant Cell* **2006**, *18*, 3548–3563. [CrossRef]

83. Krause, K.; Kilbienski, I.; Mulisch, M.; Rodiger, A.; Schafer, A.; Krupinska, K. DNA-binding proteins of the Whirly family in Arabidopsis thaliana are targeted to the organelles. *FEBS Lett.* **2005**, *579*, 3707–3712. [CrossRef]

84. Cappadocia, L.; Marechal, A.; Parent, J.S.; Lepage, E.; Sygusch, J.; Brisson, N. Crystal Structures of DNA-Whirly Complexes and Their Role in Arabidopsis Organelle Genome Repair. *Plant Cell* **2010**, *22*, 1849–1867. [CrossRef]

85. Marechal, A.; Parent, J.S.; Veronneau-Lafortune, F.; Joyeux, A.; Lang, B.F.; Brisson, N. Whirly proteins maintain plastid genome stability in Arabidopsis. *Proc. Natl. Acad. Sci. USA* **2009**, *106*, 14693–14698. [CrossRef]

86. Shedge, V.; Arrieta-Montiel, M.; Christensen, A.C.; Mackenzie, S.A. Plant mitochondrial recombination surveillance requires unusual RecA and MutS homologs. *Plant Cell* **2007**, *19*, 1251–1264. [CrossRef]

87. Khazi, F.R.; Edmondson, A.C.; Nielsen, B.L. An Arabidopsis homologue of bacterial RecA that complements an E-coli recA deletion is targeted to plant mitochondria. *Mol. Genet. Genom.* **2003**, *269*, 454–463. [CrossRef]

88. Xu, Y.Z.; Arrieta-Montiel, M.P.; Virdi, K.S.; de Paula, W.B.M.; Widhalm, J.R.; Basset, G.J.; Davila, J.I.; Elthon, T.E.; Elowsky, C.G.; Sato, S.J.; et al. MutS HOMOLOG1 Is a Nucleoid Protein That Alters Mitochondrial and Plastid Properties and Plant Response to High Light. *Plant Cell* **2011**, *23*, 3428–3441. [CrossRef]

89. Wall, M.K.; Mitchenall, L.A.; Maxwell, A. Arabidopsis thaliana DNA gyrase is targeted to chloroplasts and mitochondria. *Proc. Natl. Acad. Sci. USA* **2004**, *101*, 7821–7826. [CrossRef]

90. Morley, S.A.; Nielsen, B.L. Chloroplast DNA Copy Number Changes during Plant Development in Organelle DNA Polymerase Mutants. *Front. Plant Sci.* **2016**, *7*. [CrossRef]

91. Parent, J.S.; Lepage, E.; Brisson, N. Divergent Roles for the Two PolI-Like Organelle DNA Polymerases of Arabidopsis. *Plant Physiol.* **2011**, *156*, 254–262. [CrossRef]

92. Trasvina-Arenas, C.H.; Baruch-Torres, N.; Cordoba-Andrade, F.J.; Ayala-Garcia, V.M.; Garcia-Medel, P.L.; Diaz-Quezada, C.; Peralta-Castro, A.; Ordaz-Ortiz, J.J.; Brieba, L.G. Identification of a unique insertion in plant organellar DNA polymerases responsible for 5′-dRP lyase and strand-displacement activities: Implications for Base Excision Repair. *DNA Repair* **2018**, *65*, 1–10. [CrossRef]

93. Ayala-Garcia, V.M.; Baruch-Torres, N.; Garcia-Medel, P.; Brieba, L.G. Plant organellar DNA polymerases paralogs exhibit dissimilar nucleotide incorporation fidelity. *FEBS J.* **2018**. [CrossRef]

94. Moriyama, T.; Sato, N. Enzymes involved in organellar DNA replication in photosynthetic eukaryotes. *Front. Plant Sci.* **2014**, *5*. [CrossRef]

95. Eun, H.-M. 6—DNA Polymerases. In *Enzymology Primer for Recombinant DNA Technology*; Eun, H.-M., Ed.; Academic Press: San Diego, CA, USA, 1996; pp. 345–489.

96. Bedford, E.; Tabor, S.; Richardson, C.C. The thioredoxin binding domain of bacteriophage T7 DNA polymerase confers processivity on Escherichia coli DNA polymerase I. *Proc. Natl. Acad. Sci. USA* **1997**, *94*, 479–484. [CrossRef]

97. McCulloch, S.D.; Kunkel, T.A. The fidelity of DNA synthesis by eukaryotic replicative and translesion synthesis polymerases. *Cell Res.* **2008**, *18*, 148–161. [CrossRef]

98. Baruch-Torres, N.; Brieba, L.G. Plant organellar DNA polymerases are replicative and translesion DNA synthesis polymerases. *Nucleic Acids Res.* **2017**, *45*, 10751–10763. [CrossRef]

99. Peralta-Castro, A.; Baruch-Torres, N.; Brieba, L.G. Plant organellar DNA primase-helicase synthesizes RNA primers for organellar DNA polymerases using a unique recognition sequence. *Nucleic Acids Res.* **2017**, *45*, 10764–10774. [CrossRef]

100. Spelbrink, J.N.; Li, F.Y.; Tiranti, V.; Nikali, K.; Yuan, Q.P.; Tariq, M.; Wanrooij, S.; Garrido, N.; Comi, G.; Morandi, L.; et al. Human mitochondrial DNA deletions associated with mutations in the gene encoding Twinkle, a phage T7 gene 4-like protein localized in mitochondria. *Nat. Genet.* **2001**, *29*, 223–231. [CrossRef]

101. Korhonen, J.A.; Gaspari, M.; Falkenberg, M. TWINKLE has 5'-> 3' DNA helicase activity and is specifically stimulated by mitochondrial single-stranded DNA-binding protein. *J. Biol. Chem.* **2003**, *278*, 48627–48632. [CrossRef]

102. Levikova, M.; Cejka, P. The Saccharomyces cerevisiae Dna2 can function as a sole nuclease in the processing of Okazaki fragments in DNA replication. *Nucleic Acids Res.* **2015**, *43*, 7888–7897. [CrossRef]

103. Li, Z.; Liu, B.; Jin, W.; Wu, X.; Zhou, M.; Liu, V.Z.; Goel, A.; Shen, Z.; Zheng, L.; Shen, B. hDNA2 nuclease/helicase promotes centromeric DNA replication and genome stability. *EMBO J.* **2018**, *37*. [CrossRef]

104. Jia, N.; Liu, X.M.; Gao, H.B. A DNA2 Homolog Is Required for DNA Damage Repair, Cell Cycle Regulation, and Meristem Maintenance in Plants. *Plant Physiol.* **2016**, *171*, 318–333. [CrossRef]

105. Cho, H.S.; Lee, S.S.; Kim, K.D.; Hwang, I.; Lim, J.-S.; Park, Y.-I.; Pai, H.-S. DNA gyrase is involved in chloroplast nucleoid partitioning. *Plant Cell* **2004**, *16*, 2665–2682. [CrossRef]

106. Fuste, J.M.; Wanrooij, S.; Jemt, E.; Granycome, C.E.; Cluett, T.J.; Shi, Y.H.; Atanassova, N.; Holt, I.J.; Gustafsson, C.M.; Falkenberg, M. Mitochondrial RNA Polymerase Is Needed for Activation of the Origin of Light-Strand DNA Replication. *Mol. Cell* **2010**, *37*, 67–78. [CrossRef]

107. Ramachandran, A.; Nandakumar, D.; Deshpande, A.P.; Lucas, T.P.; R-Bhojappa, R.; Tang, G.Q.; Raney, K.; Yin, Y.W.; Patel, S.S. The Yeast Mitochondrial RNA Polymerase and Transcription Factor Complex Catalyzes Efficient Priming of DNA Synthesis on Single-stranded DNA. *J. Biol. Chem.* **2016**, *291*, 16828–16839. [CrossRef]

108. Smith, D.R. Updating Our View of Organelle Genome Nucleotide Landscape. *Front. Genet.* **2012**, *3*, 175. [CrossRef]

109. Larson, M.A.; Bressani, R.; Sayood, K.; Corn, J.E.; Berger, J.M.; Griep, M.A.; Hinrichs, S.H. Hyperthermophilic Aquifex aeolicus initiates primer synthesis on a limited set of trinucleotides comprised of cytosines and guanines. *Nucleic Acids Res.* **2008**, *36*, 5260–5269. [CrossRef]

110. Arnold, J.J.; Smidansky, E.D.; Moustafa, I.M.; Cameron, C.E. Human mitochondrial RNA polymerase: Structure-function, mechanism and inhibition. *Biochim. Biophys. Acta* **2012**, *1819*, 948–960. [CrossRef]

111. Hess, W.R.; Borner, T. Organellar RNA polymerases of higher plants. *Int. Rev. Cytol.* **1999**, *190*, 1–59. [CrossRef]

112. Yin, C.; Richter, U.; Borner, T.; Weihe, A. Evolution of plant phage-type RNA polymerases: The genome of the basal angiosperm Nuphar advena encodes two mitochondrial and one plastid phage-type RNA polymerases. *BMC Evol. Biol.* **2010**, *10*, 379. [CrossRef]

113. Baba, K.; Schmidt, J.; Espinosa-Ruiz, A.; Villarejo, A.; Shiina, T.; Gardestrom, P.; Sane, A.P.; Bhalerao, R.P. Organellar gene transcription and early seedling development are affected in the rpoT; 2 mutant of Arabidopsis. *Plant J.* **2004**, *38*, 38–48. [CrossRef]

114. Serino, G.; Maliga, P. RNA polymerase subunits encoded by the plastid rpo genes are not shared with the nucleus-encoded plastid enzyme. *Plant Physiol.* **1998**, *117*, 1165–1170. [CrossRef]

115. Nielsen, B.L.; Rajasekhar, V.K.; Tewari, K.K. Pea Chloroplast DNA Primase—Characterization and Role in Initiation of Replication. *Plant Mol. Biol.* **1991**, *16*, 1019–1034. [CrossRef]

116. Morley, S.A.; Peralta-Castro, A.; Brieba, L.G.; Miller, J.; Ong, K.L.; Ridge, P.G.; Oliphant, A.; Aldous, S.; Nielsen, B.L. Arabidopsis thaliana organelles mimic the T7 phage DNA replisome with specific interactions between Twinkle protein and DNA polymerases Pol1A and Pol1B. *BMC Plant Biol.* **2019**, *19*, 241. [CrossRef]

117. Desveaux, D.; Marechal, A.; Brisson, N. Whirly transcription factors: Defense gene regulation and beyond. *Trends Plant Sci.* **2005**, *10*, 95–102. [CrossRef]

118. Redei, G.P. Extrachromosomal mutability determined by a nuclear gene locus in Arabidopsis. *Mutat. Res.* **1973**, *18*, 149–162. [CrossRef]

119. Martinezzapater, J.M.; Gil, P.; Capel, J.; Somerville, C.R. Mutations at the Arabidopsis Chm locus promote rearrangements of the mitochondrial genome. *Plant Cell* **1992**, *4*, 889–899.

120. Abdelnoor, R.V.; Yule, R.; Elo, A.; Christensen, A.C.; Meyer-Gauen, G.; Mackenzie, S.A. Substoichiometric shifting in the plant mitochondrial genome is influenced by a gene homologous to MutS. *Proc. Natl. Acad. Sci. USA* **2003**, *100*, 5968–5973. [CrossRef]

121. Reenan, R.A.G.; Kolodner, R.D. Characterization of insertion mutations in the Saccharomyces cerevisiae Msh1 and Msh2 genes—Evidence for separate mitochondrial and nuclear Functions. *Genetics* **1992**, *132*, 975–985.

122. Abdelnoor, R.V.; Christensen, A.C.; Mohammed, S.; Munoz-Castillo, B.; Moriyama, H.; Mackenzie, S.A. Mitochondrial genome dynamics in plants and animals: Convergent gene fusions of a MutS homologue. *J. Mol. Evol.* **2006**, *63*, 165–173. [CrossRef]

123. Davila, J.I.; Arrieta-Montiel, M.P.; Wamboldt, Y.; Cao, J.; Hagmann, J.; Shedge, V.; Xu, Y.-Z.; Weigel, D.; Mackenzie, S.A. Double-strand break repair processes drive evolution of the mitochondrial genome in Arabidopsis. *BMC Biol.* **2011**, *9*, 64. [CrossRef]

124. Odahara, M.; Kishita, Y.; Sekine, Y. MSH1 maintains organelle genome stability and genetically interacts with RECA and RECG in the moss Physcomitrella patens. *Plant J.* **2017**, *91*, 455–465. [CrossRef]

125. Fukui, K.; Harada, A.; Wakamatsu, T.; Minobe, A.; Ohshita, K.; Ashiuchi, M.; Yano, T. The GIY-YIG endonuclease domain of Arabidopsis MutS homolog 1 specifically binds to branched DNA structures. *FEBS Lett.* **2018**, *592*, 4066–4077. [CrossRef]

126. Sunderland, P.A.; West, C.E.; Waterworth, W.M.; Bray, C.M. An evolutionarily conserved translation initiation mechanism regulates nuclear or mitochondrial targeting of DNA ligase 1 in Arabidopsis thaliana. *Plant J.* **2006**, *47*, 356–367. [CrossRef]

127. Waterworth, W.M.; Kozak, J.; Provost, C.M.; Bray, C.M.; Angelis, K.J.; West, C.E. DNA ligase 1 deficient plants display severe growth defects and delayed repair of both DNA single and double strand breaks. *BMC Plant Biol.* **2009**, *9*. [CrossRef]

 *plants*

Review

Structure–Function Analysis Reveals the Singularity of Plant Mitochondrial DNA Replication Components: A Mosaic and Redundant System

Luis Gabriel Brieba

Laboratorio Nacional de Genómica para la Biodiversidad, Centro de Investigación y de Estudios Avanzados del IPN, Apartado Postal 629, Irapuato, Guanajuato C.P. 36821, Mexico; luis.brieba@cinvestav.mx

Received: 24 October 2019; Accepted: 19 November 2019; Published: 21 November 2019

Abstract: Plants are sessile organisms, and their DNA is particularly exposed to damaging agents. The integrity of plant mitochondrial and plastid genomes is necessary for cell survival. During evolution, plants have evolved mechanisms to replicate their mitochondrial genomes while minimizing the effects of DNA damaging agents. The recombinogenic character of plant mitochondrial DNA, absence of defined origins of replication, and its linear structure suggest that mitochondrial DNA replication is achieved by a recombination-dependent replication mechanism. Here, I review the mitochondrial proteins possibly involved in mitochondrial DNA replication from a structural point of view. A revision of these proteins supports the idea that mitochondrial DNA replication could be replicated by several processes. The analysis indicates that DNA replication in plant mitochondria could be achieved by a recombination-dependent replication mechanism, but also by a replisome in which primers are synthesized by three different enzymes: Mitochondrial RNA polymerase, Primase-Helicase, and Primase-Polymerase. The recombination-dependent replication model and primers synthesized by the Primase-Polymerase may be responsible for the presence of genomic rearrangements in plant mitochondria.

Keywords: DNA replication; evolution; replisome; recombination-dependent replication

1. Introduction

1.1. Plant Mitochondria Genomes

Mitochondria arose from a monophyletic endosymbiotic event between an archaea and an α-proteobacteria approximately two billion years ago [1]. During the evolution of eukaryotes, mitochondrial genomes have evolved in size and complexity. For instance, mitochondrial genomes vary in size more than three orders of magnitude and they exist as circular, linear, linear-branched, linear-fragmented, and mixtures of maxi and mini-circles [2]. In general, metazoan mitochondrial genomes are circular molecules that vary in sizes between 10 to 30 kb [3]. In contrast, plant mitochondrial genomes are predominantly large linear DNA molecules (up to 11 Mb in angiosperms from the genus Silene). Besides the differences between the physical structure of the plant and metazoan genomes (linear versus circular), the most remarkable characteristics of plant mitochondrial genomes are their ability to rearrange, their low nucleotide substitution rate, and the evolution of new mitochondrial open reading frames. For instance, almost all vertebrates exhibit a similar organization in their mitochondrial genome arrangement [4], whereas the mitochondrial genomic organization in plants is different even between ecotytpes of the same species [5]. The abundance of noncoding sequences severely complicates alignments of mitochondrial genomes from different plant families [6]. A comparison between the mitochondrial genomes of Col-0 and C24 ecotypes of *Arabidopsis thaliana*, that diverged 200,000 years ago, shows that both genomes exhibit different configurations because of a large inverted repeat [5,7–9].

Even though plant mitochondrial genomes rearrange, the substitution rate in their coding regions is almost negligible, in contrast with the highly mutable human mitochondrial genome [10,11].

1.2. Replication in Mammalian Mitochondria

Due to their bacterial origin, the mechanisms involved in mitochondrial and plastid DNA replication are expected to be related to bacteria. Yet mitochondrial DNA replication in metazoans is achieved by a replisome that is phylogenetically related to the bacteriophage T7 replisome [12,13]. In mitochondrial replisomes from metazoans, a bacteriophage-related RNA polymerase synthesizes RNA primers to start replication at the heavy and light chains of the circular DNA mitochondrial molecule, a hexameric helicase unwinds double-stranded DNA, and a trailing mitochondrial DNA polymerase synthesizes DNA. Human mitochondrial DNA replication starts by a strand-displacement model of replication in which human mitochondrial RNA polymerase (RNAP) transcribes the heavy-strand promoter generating a primer that is processed and passed on to the mitochondrial DNA polymerase (DNAP), DNA replication proceeds interruptedly to copy a new heavy-strand [14]. During this process, the replication fork replicates the light strand origin of replication. This DNA sequence folds into a stem–loop structure that allows primer synthesis by the mitochondrial RNAP, and these primers are elongated by the mitochondrial DNA polymerase [15]. Elongation of the heavy and light chains continues asynchronically until the two chains are completely copied. Although the strand-displacement model is generally accepted as the mechanism for mitochondrial DNA replication, there are discrepancies regarding how it proceeds. To date, two alternative models explain strand-asynchronous replication in mitochondria. One model proposes that long RNA molecules hybridize to the single-stranded heavy-strand [16]. This ribonucleotide (RNA) incorporation occurred throughout the lagging strand (RITOLS) transcripts that are continuously hybridized as replication continues [17]. The second model proposes that single-stranded DNA binding proteins coat the lagging-strand template [18]. Alternatively to the strand-displacement model, coupled leading and lagging-strand DNA synthesis can occur bidirectionally in mitochondria [19,20] and recent work stablished that cells can shift between the strand-asynchronous and the coupled leading and lagging-strand DNA synthesis depending of the amount of transcripts [21].

2. Enzymes Involved in Organelle DNA Replication in Plants Can Be Grouped into Bacteriophage-Related, Replication-Dependent Replication and Unique Enzymes

The main difference between the mitochondrial metazoan and bacteriophage T7 replisomes is that the T7 primase-helicase harbors an active primase module that synthesizes primers for lagging strand synthesis, whereas the primase module of metazoan primase-helicases is inactive and primer synthesis depends solely on the mitochondrial RNA polymerase [22,23]. Thus, metazoan primase-helicases harbors a primase module that has lost its priming activities. The similarities between the metazoan replicative mitochondrial DNA primase-helicase and the primase-helicase of bacteriophage T7 resulted in the name of TWINKLE (T7 gp4-like protein with intra-mitochondrial nucleoid localization) for this protein [24].

2.1. A T7-Like Replisome in Plant Organelles

In this review, we focus on the proteins from the model plant *Arabidopsis thaliana* as a representative of flowering plants. As their metazoan counterparts, plant organelles harbor enzymes related to the T7 replisome (Table 1). From the four enzymes involved in DNA replication in bacteriophage T7 and metazoan mitochondria, land plants have conserved three of them: (a) The primase-helicase, (b) the RNA polymerase, and (c) the single-stranded DNA binding protein (Table 1). The presence of these proteins suggests that plant mitochondrial DNA replication is executed in part by a mechanism that resembles the coordinated leading and lagging-strand replication model of bacteriophage T7 [22]. In this model, a central primase-helicase unwinds dsDNA in the 3′-5′direction followed by a processive DNA polymerase in the leading strand. The primase module of the primase-helicase uses the unwounded single-stranded regions to recognize a sequence to start the synthesis of very short ribonucleotides that are handed off to the active site of the lagging strand DNA polymerase. The single-stranded DNA regions generated during this trombone mechanism are coated by the single-stranded binding proteins [22,25].

Table 1. Proteins related to bacteriophage T7 proteins present in plant mitochondria.

Enzyme	Phage T7	Human Mitochondria	Arabidopsis Organelles	Number	Localization
DNA Polymerase	T7 DNAP	DNAPγ	—	—	—
Helicase-Primase	Helicase-Primase	Helicase	AtTwinkle	At1g30680	Chloroplast and mitochondria
Primase	—	—	AtTwinky	At1g30660	?
Primase	T7 RNAP	mtRNAP	RpoTm	At1g68990	Mitochondria
—	—	—	RpoTmp	At5g15700	Chloroplast and mitochondria
SSB	SSB	mtSSB	mtAtSSB1	At4g11060	Chloroplast and mitochondria
—	—	—	mtAtSSB2	At3g18580	Mitochondria

2.1.1. Plant Organellar Primase-Helicase (AtTwinkle)

Primase-helicases are the central component of replisomes [26,27]. These enzymes unwind double-stranded DNA segments using NTP hydrolysis for translocation and primer synthesis, using their helicase and primase modules, respectively [22,27]. The organellar primase-helicases in *A. thaliana* (dubbed AtTwinkle) is a 709 amino acid protein with mitochondria and chloroplast localization [28] (Figure 1A). AtTwinkle, as predicted for all plant primase-helicases, harbors both primase and helicase activities [28–30]. Structural studies of primase-helicase show that these enzymes assemble as heptamers or hexamers in which the helicase modules form a compact oligomeric ring to which the primase modules attach [31,32] (Figure 1B,C). The primase module of AtTwinkle contains six conserved motifs [30]. Motif I corresponds to the zinc binding domain (ZBD) necessary for template recognition, whereas regions II to VI assemble the RNA Polymerase domain (Figure 1D). In contrast to all previously characterized primase-helicases, AtTwinkle recognizes two cryptic nucleotides within the ssDNA template [29], a biochemical property that may reduce the length of the Okazaki fragments during plant mitochondrial replication. The helicase module of AtTwinkle shares high amino acid identity with the helicase module of the T7 primase-helicase and harbors the five conserved motifs [33], including a Walker motif necessary for nucleotide hydrolysis. The presence of an active AtTwinkle protein in Arabidopsis suggests the presence of a plant mitochondrial replisome in which a DNA polymerase replicates DNA following the unwinding of the double helix and exposing the leading-strand for continuous synthesis [34]. The primase activity suggests that a trailing DNA polymerase synthesizes the lagging-strand using primers synthesized by the primase module of AtTwinkle [29]. This model of coordinated leading and lagging strands occurs in bacteriophages T4 and T7, but not in mitochondria from metazoans and yeast [22,23,26]. Interestingly, Arabidopsis harbors a protein that contains the zinc finger and the RNA polymerase module of AtTwinkle dubbed AtTwinky [28]. This module by itself is functional in vitro [29]. An Arabidopsis insertional line in AtTwinkle shows no apparent phenotype, maybe because the T-DNA insertion occurs in an intron or because of redundant mechanisms for primer synthesis and DNA unwinding [34].

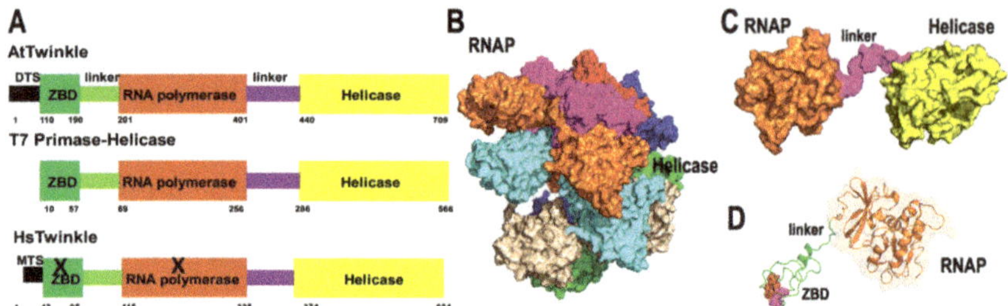

Figure 1. AtTwinkle is a homolog of bacteriophage T7 primase-helicase and mitochondrial Twinkle. (**A**) Schematic representation of the bifunctional T7 primase-helicase in comparison to AtTwinkle and human Twinkle. T7 primase-helicase and AtTwinkle contain the conserved motifs necessary for primase and helicase activities, whereas human Twinkle is inactive as a primase. (**B**) Homology model of AtTwinkle showing its RNA polymerase domain and helicase modules with basis on the crystal structure of the heptameric T7 primase-helicase [31]. (**C**) Close view of a monomeric module of the RNAP and helicase of AtTwinkle. (**D**) Close view of the primase module composed of the zinc binding domain (ZBD) and RNAP domain. The conserved cysteines that coordinate the zinc atom are colored in red and magenta.

2.1.2. Bacteriophage-Type Plant Organellar RNA Polymerases

In yeast and metazoans mitochondria, transcription is carried out by a single RNA polymerase (mtRNA) homologous to T7 RNA polymerase [35]. In contrast to metazoans that harbor one nuclear-encoded mtRNAP, flowering plants encode three bacteriophage-type RNA polymerases [36,37]. One is localized into the mitochondria (RpoTm), one into the chloroplast (RpoTp), and the third one presents dual mitochondrial and plastid localization (RpoTmp). In Arabidopsis, RpoTm and RpoTp start transcription at a specific set of promoters. However, RpoTmp is unable to start transcription by itself [38]. These enzymes are closely related to bacteriophage T7 RNAP and due to sequence similarity are expected to fold into two conserved domains: An N-terminal domain, possibly involved in RNA binding and a C-terminal or polymerization domain. The C-terminal domain is structurally divided into three subdomains, dubbed palm, fingers, and thumb (Figure 2). Yeast and metazoan mitochondrial RNAPs are only active by themselves on supercoiled templates; on linearized templates, they need an associated transcription factor to start transcription [39,40]. Likewise, plant mitochondrial RNAPs are only active in supercoiled templates [36], suggesting that they also need an unidentified plant mitochondrial transcription factor for efficient promoter melting. Mitochondrial RNAPs from metazoans and yeast contains an N-terminal pentatricopeptide repeat (PPR) not present in plant mitochondrial RNAPs and T7 RNAP (Figure 2). Thus, plant mitochondrial RNAP are more compact than yeast and metazoan mitochondrial RNAPs.

In bacteriophage T7 and metazoan mitochondria, their RNAPs synthesize long RNA chains at defined sequences that mark their origins of replication [15,41–43]. It is unknown if plant mitochondrial RNAPs play a role in synthesizing RNA primers during mitochondrial or plastid replication. However, plant mitochondrial genomes are proposed to exist as a multitude of linear fragments, carrying only partial segments of their genome [44–46]. The presence of numerous promoter DNA sequences in plant mitochondria makes possible the existence of multiple initiation replication sites in mitochondrial DNA.

During metazoan mitochondrial DNA replication, the RNA primers generated by the mitochondrial RNA polymerase are removed by a specific set of nucleases. In humans, five different nucleases participate in this process [47–52]. From those enzymes, RNAse H1 plays a predominant role by degrading the RNA primer until it reaches few nucleotides. These last two to three ribonucleotides can be removed by the flap specific nucleases FEN1, DNA2, and MGME1 or by the selective 5′-3′ exonuclease EXOG [48,51]. Arabidopsis encodes for three proteins highly homologous to RNase H1, dubbed AtRNH1A (At3g01410), AtRNH1B(At5g51080), and AtRNH1C (At1g24090) [47]. AtRNH1A is localized into the nucleus, whereas AtRNH1B and AtRNH1C are imported into mitochondria and chloroplasts, respectively. AtRNH1C prevents R-loop accumulation in chloroplast especially at highly transcribed regions and putative origins of replication [47]. AtRNH1C is involved in assuring genome stability in the chloroplast, suggesting the possibility that AtRNH1B may contribute to the removal of RNA primers in plant mitochondria.

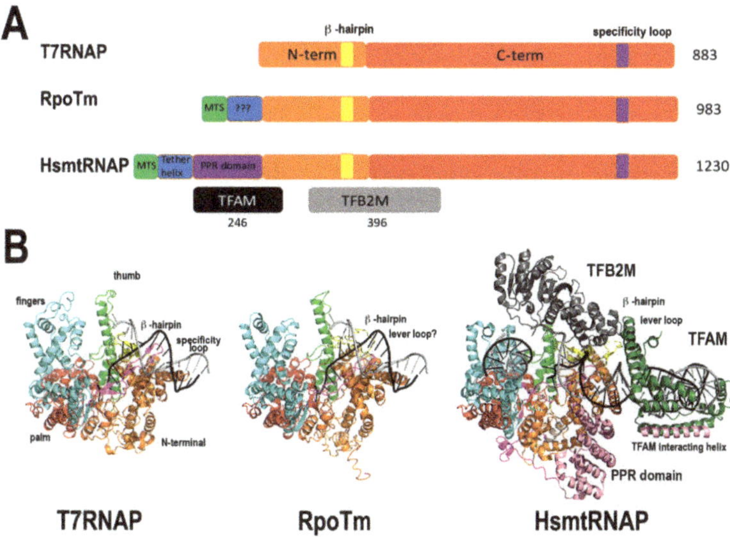

Figure 2. Bacteriophage-type plant organellar RNA polymerases. (**A**) Domain organization of bacteriophage-related RNAP. These enzymes share a C-terminal or polymerization domain that is divided into three subdomains: Fingers, palm, and thumb, and a N-terminal domain involved in promoter opening and RNA binding. The N-terminal domain is colored orange and the subdomains of the fingers, thumb, and palm of blue, green, and red, respectively. mtHsRNAP associates with two accessory subunits (TFB2M and TFAM) to open double-stranded DNA and contains a N-terminal pentatricopeptide repeat (PPR)-domain and a tether helix not present in plant mitochondrial RNAPs. (**B**) Structural model of the mtAtRNAP compared to bacteriophage T7RNAP and human mtRNAP during transcription initiation [40,53].

2.1.3. Plant Organellar Single-Stranded DNA Binding Proteins

All replisomes contain single-stranded DNA binding proteins (SSBs) that coat the lagging-strand DNA chain and exert a multitude of interactions with DNA polymerases, DNA helicases, and other proteins involved in DNA metabolism. Flowering plants encode for two canonical single-stranded DNA binding proteins that are targeted to mitochondria (AtmtSSB1 and AtmtSSB2) [54,55]. Like all SSBs, these proteins harbor an oligonucleotide/oligosaccharide/binding (OB)-fold domain and share a conserved set of aromatic amino acids that in other bacterial and mitochondrial SSBs are important for binding to single-stranded DNA. Among these amino acids, residues W54 and F60 that are determinant for binding to SSB in bacteria are conserved in AtmtSSB1 and AtmtSSB2 [56–58] (Figure 3). AtmtSSB1 assembles as a tetramer, binds single-stranded DNA in the nanomolar range, and interacts with plant mitochondrial DNA polymerases from Arabidopsis [59]. A recent proteomic analysis indicates that both AtmtSSB1 and AtmtSSB2 are highly abundant proteins, suggesting that a great portion of the mitochondrial single-stranded DNA is coated with them [55]. The last nine amino acids of *E. coli* SSB are responsible for mediating protein–protein interactions [60,61]. AtmtSSB1 contains a predominant acid tail while AtmtSSB2 harbors an aromatic tail (Figure 3), suggesting the possibility that both SSBs exert differential protein–protein interactions.

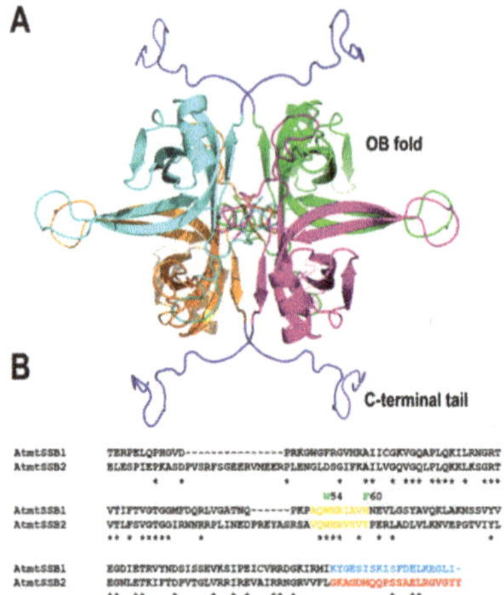

Figure 3. Homology model of tetrameric AtmtSSB1. (**A**) Homology model of AtmtSBB1 illustrating its oligonucleotide/oligosaccharide/binding (OB)-fold and an acid C-terminal tail. (**B**) An amino acid sequence alignment illustrates that the C-terminal tail of AtmtSSB2 is composed of two aromatic amino acids, whereas AtmtSSB1 is acidic.

2.2. A Putative Recombination-Dependent Replication System in Plant Mitochondria

One of the main differences between plant and human mitochondrial genomes resides in the presence of highly abundant repeats of different lengths in plant mitochondria [62,63]. These repeats are classified by Gualberto and Newton as large repeats (>500 base pairs); intermediate-sized repeats (50–500 base pairs); and small repeats (<50 base pairs) [64,65]. Seminal studies deduced that the recombinogenic character at large repeats is responsible for plant mitochondrial DNA genomic configurations [62,66,67]. Thus, it is generally accepted that recombination at large repeats results in the presence of multiple mitochondrial genome conformations, whereas recombination at intermediate-size repeats are not as frequent [5,68]. The low-frequency recombination at intermediate-size repeats leads to changes in the stoichiometry of the mitochondrial genomes [69,70]. Finally, recombination at small repeats drives the apparition of new open reading frames associated with traits like cytoplasmic male sterility [71,72]. The notion that recombination is dependent on the length of the repeat is challenged by comparing new mitochondrial DNA sequences between domesticated and wild-type cultivars and by following the evolutionary history between species [73,74].

The recombinant character of the mitochondrial genome is reminiscent of bacteriophage T4 genome, which uses a recombination-dependent replication (RDR) mechanism [46,75]. Furthermore, seminal studies have shown the presence of linear molecules, head-to-tail concatemers, branched, and rosette-like structures during plant mitochondrial replication suggesting that free single-stranded DNA ends direct primer formation [45,46,76,77]. In contrast to metazoan mitochondria, plant mitochondria harbor a complete set of enzymes involved in HR. In bacteriophage, T4 RDR starts by coating of the single-stranded DNA by a recombinase dubbed UvsX, a protein homolog to bacterial RecA, or eukaryotic Rad51. As all recombinases, this protein uses ATP to catalyze the exchange of the single-stranded DNA into double-stranded DNA. This initial step creates a triple-stranded DNA region in which T4 DNA polymerase assembles to initiate replication. A replicative helicase loads onto the displaced DNA strand, this enzyme translocates in 5′ to 3′ direction, unwinding DNA, and generating a template for the trailing polymerase. The helicase associates with a primase that recognizes single-stranded sequences in the 3′-5′direction and generates primers used by a second DNA polymerase during replisome assembly. Although this system is relatively simple, it needs the presence of several mediator proteins that coordinate protein loading. In Arabidopsis mitochondria, several homologs to the battery of T4 enzymes involved in RDR are present, suggesting the possibility that RDR is a functional mechanism in plants (Table 2).

Table 2. Plant mitochondrial proteins related to bacteriophage T4 recombination-dependent replication proteins.

Process or Enzyme	Phage T4	Bacteria	Arabidopsis Organelles	Acession Number	Localization
Annealing to ssDNA	UvsX	RecA	AtRecA2	At2g19490	Chloroplast and mitochondria
			AtRecA3	At3g10140	Mitochondria
Suppress Rec-A Annealing	—	RecX	AtRecX	At3g13226.1	?
Mediator	UvsY	RecFOR	AtODB1	At1g71310	Mitochondria and the nucleus
Helicase	Dda?	RadA	AtRadA	At5g50340	Chloroplast and mitochondria
Branch Migration Remodeling	UvsW	RecG	AtRecG1	At2g01440.1	RecG1

2.2.1. AtRecA

RecA and its homologs Rad51 and BRCA are the central components of homologous recombination. RecA is an archetypical bacterial recombinase that loads onto resected single-stranded DNA in an ATP-dependent reaction. It assembles a nucleic acid-protein filament that navigates the double-stranded genome in search of a homologous sequence, and when a region of homology is encountered, this filament perfectly pairs with its homologous partner (located within a dsDNA region) and generates a heteroduplex or D-loop intermediate [78]. HR by Rad51/RecA is abrogated in the presence of mismatches and bacterial RecA needs at least eight nucleotides of perfect complementarity to form a stable D-loop, although the efficiency of heteroduplex formation increases according to the length of the perfect complementarity [79–81]. In bacteria, the RecA monomer consists of a central or core domain of approximately 230 amino acids. This domain folds into a single β-sheet and six α-helices [82]. This core domain is flanked by N and C-terminal domains of approximately 30 and 60 amino acids, respectively [82]. The crystal structure of bacterial RecA–ssDNA filament illustrates how the RecA assembles onto ssDNA and how Watson–Crick pairing is assured during the homology search [83] (Figure 4A).

Unlike metazoan mitochondria that are devoid of RecA homologs, plant mitochondria harbor orthologues of the recombinase RecA/Rad51 gene family [69,84–86]. These proteins are conserved from algae to flowering plants. Genetic studies in *Physcomitrella* patens and *Arabidopsis* demonstrate the role of RecA in preventing illegitimate recombination events at small repeats in *P. patents* and intermediate-size repeats in Arabidopsis [69,86,87]. *A. thaliana* harbors three RecA genes. RecA1 is targeted to the chloroplast, RecA2 is targeted to plastids and mitochondria, whereas RecA3 is only targeted to mitochondria [69,88]. AtRecA1 is an essential gene, whereas AtRecA2 is only necessary after the seedling stage [69,86]. AtRecA2 and AtRecA3 share 53% and 41% amino acid identity with *E. coli* RecA, respectively. The latter suggests that HR in plant mitochondria may follow a mechanism similar to bacteria. Interestingly, AtRecA3 lacks the last 22 amino acids of its C-terminal domain in comparison to *E. coli* RecA. In bacteria, these residues have a highly acidic composition and a deletion of 17 amino acids is more efficient in displacing bacterial SSB from ssDNA, thus the C-terminal extension negatively modulates RecA activity [89] (Figure 4). Plants mutated in AtRecA3 are phenotypically normal. However, they are sensitive to genotoxic treatments [69]. The loss of RecA2 and RecA3 promotes rearrangements at intermediate-size repeats [86]. These repeats are not perfect and lead to homeologous recombinant products (illegitimate recombination products). The increase of illegitimate recombination products in the absence of AtRecA2 or AtRecA3 suggests that less stringent RecA-independent pathways take over in their absence. One possible pathway is the single-strand annealing recombination pathway (SSA) under the control of specialized SSBs with annealing capabilities as is the case in *Deinococcus radiodurans* [90]. Recent proteomic studies indicate that RecA2 is one of the most abundant DNA binding proteins in plant mitochondria [55].

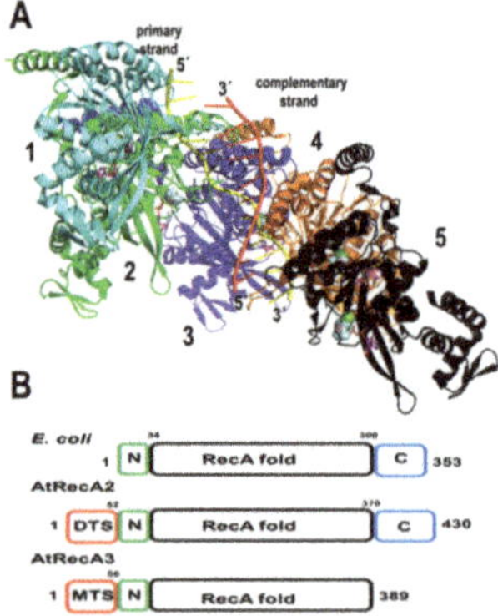

Figure 4. Structural conservation of plant and bacterial RecAs. (**A**) Crystal structure of the bacterial RecA postsynaptic nucleoprotein filament determined by Chen, Yang, and Pavletich [83]. Each of the five RecA monomers is individually colored and labeled with numbers. The search strand is colored in yellow and the complementary strand in red. The crystal structure comprises solely the RecA fold and the C-terminal domain is not present in the initial construct. (**B**) Domain organization of AtRecA2 and AtRecA3 in comparison to bacterial RecA. AtRecA3 lacks the C-terminal regulatory domain.

2.2.2. AtRecX

In bacteria, RecA can be inhibited by an interaction with a small protein (approximately 20 kDa) dubbed RecX [91]. RecX proteins bind to RecA monomers and DNA [92]. Bacterial RecX proteins are composed of nine α-helices that arrange into three three-helix bundles [93,94] (Figure 5A). RecX binds to RecA filaments promoting their dissociation from single-stranded DNA and impinging homologous recombination [95,96]. *A. thaliana* encodes for a gene of 382 amino acids, ortholog to bacterial RecX, with a predicted mitochondrial localization signal in its first 25 amino acids, a domain of unknown function and a C-terminal segment that presents 30% amino acid identity with *E. coli* RecX (Figure 5B). The presence of this RecX ortholog (AtRecX) suggests the possibility that RecA activities are subject to regulation in plants. The presence of three RecA genes in flowering plants also suggests that these proteins may be subject to a gradient of regulation by RecX in vivo. In the moss *Physcomitrella patens* RECX, overexpressing mutants exhibit increased recombination products at short dispersed repeats in mitochondria [97], suggesting that RecX modulates RecA activity and when RecA is not functionally active, less accurate DNA repair routes gain access to ssDNA with a concomitant appearance of illegitimate recombination products.

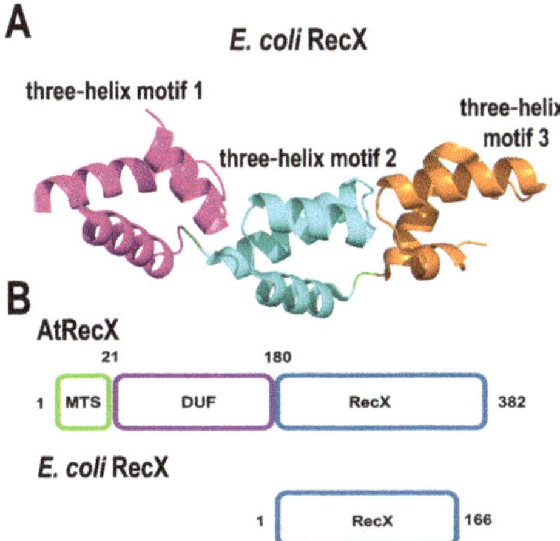

Figure 5. Structural organization of AtRecX. (**A**) Crystal structure of RecX from *E. coli* (PDB: 3c1d). RecX is composed of three repeats of a three-helix motifs, (**B**) modular organization of AtRecX in comparison to bacterial RecX. Plant RecX harbor a mitochondrial targeting sequence (MTS) and a N-terminal domain of unkown function. AtRecX share more than 30% amino acid identity with bacterial RecXs.

2.2.3. Organellar DNA-Binding Proteins (ODBs)

Upon the formation of single-stranded breaks, canonical SSBs bind to ssDNA blocking its acess to other binding proteins. In order for RecA to bind ssDNA, SSBs have to be removed from ssDNA. In bacteria, a protein named RecO (or its functional homolog in yeast, Rad52) interacts with the C-terminal tails of SSBs creating space for RecA binding [98]. Via proteomic studies, the Gualberto group identified that Arabidopsis contains two organellar DNA-binding proteins (ODBs), one located in the mitochondria (AtODB1) and the other in the chloroplast (AtODB2) [99]. AtODBs are homologous to Rad52 and the yeast mitochondrial nucleoid protein Mgm101 [100]. Mgm101 assembles an oligomeric ring structure and preferentially binds single-stranded DNA, suggesting a role in stabilizing and annealing DNA segments [101,102]. Likewise, Rad52 induces the displacement of human replication protein A (RPA) from ssDNA, anneals complementary ssDNA strands, and promotes strand exchange between ssDNA and dsDNA [103]. Thus, Rad52 promotes HR by displacing RPA, and promotes the coating of Rad51 by directing single-stranded annealing. Crystal structures of human Rad52 in complex with ssDNA depict this molecule as an undecameric ring in which two Rad52 oligomers could mediate HR in trans [104–106]. AtODB1 comprises 177 amino acids and shares extensive homology with the N-terminal domain of Rad52 (that contains the DNA binding and oligomerization regions). However, AtODB1 lacks a C-terminal domain containing the interacting motif for RPA and Rad51, that are involved in their displacement from ssDNA [107,108]. AtODB1 is 41 amino acids shorter than the construct of 212 amino acids used to crystallize human Rad52. Interestingly, the last 41 amino acids of human Rad52 folds into an alpha-helix (named helix 5) that intercalates with the first alpha-helix of the structure stabilizing the oligomeric assembly [104] (Figure 6).

Because of the reduced size of AtODBs, it is unknown if these proteins interact with SSBs from plant mitochondria like AtmtSSBs, AtWhirlies, AtRecA, or AtOSBs. Arabidopsis odb1 insertional mutants present no variation in phenotype, however upon genotoxic stress, they show inferior homologous recombination potential and increased microhomology-mediated end joining (MMEJ) [100]. This suggests that plant ODBs may function as mediator proteins that promote the annealing of plant RecAs onto single-stranded DNA. Recombinantly expressed plant ODB1 can anneal short DNA sequences [100]. The increase in MMEJ in plants lacking AtODB1 may be related to a role of this protein in a single-strand annealing recombination pathway, since human Rad52 proteins promote this route [109,110].

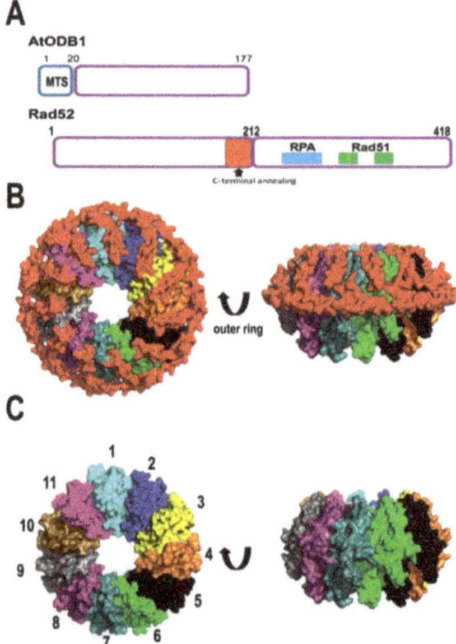

Figure 6. AtODB1 resembles human Rad52. (**A**) Structural domain organization of AtODB1 in comparison to human Rad52. AtODB1 lacks the C-terminal domain necessary to interact with RPA and Rad51; (**B**) crystal structure of the undecameric ring of human Rad52. The undecameric structure is stabilized by alpha-helix 5 that interacts with alpha-helix 1 of the neighbor molecule. Each subunit (residues 1 to 172 is individually colored) and the C-terminal residues (172 to 212) are colored in read. (**C**) Model of AtODB1 as a undecameric ring lacking alpha-helix 5 of human Rad52.

2.2.4. AtRadA

Bacterial RadA promotes single-stranded strand exchange similar to RecA, and was initially suggested to be orthologous to RecA [111]. Bacterial RadAs have a conserved domain organization composed of: (a) A putative zinc finger (ZnF), (b) a Rec-A like ATPase domain with a unique KNRFG motif, and (c) a region homologous to the Lon protease. Gualberto and Newton have identified the presence of a RadA-like gene in plant organelles [64] (At5g50340.1). This protein harbors a dual organellar targeting sequence in its first 88 amino acids and has 63% amino acid similarity with RadA from *Streptococcus pneumoniae* [112–114]. Bacterial Rad assembles as a hexameric ring, resembling the structural organization of replicative DnaB helicases [112] (Figure 7). Bacterial RadA interacts with RecA and unwinds dsDNA in the 3′-5′ direction. These biochemical properties suggest that RadA promotes the extension of ssDNA after RecA mediated homologous recombination, similar to the extension of bacterial origins of replication mediated by DnaB [112].

Because of the conserved domain organization of AtRadA, it is plausible that this protein is involved in a recombination-dependent replication mechanism. The appearance of multiple origins of replication in plant mitochondria by electron microscopy suggests the possibility that the unwinding ability of AtRadA is a key element for break-induced replication, by stabilizing a D-loop in synchrony with AtRecAs in which AtPolIs could be loaded. An interaction between RecA and RadA promotes D-loop extension in bacteria [115], suggesting that a similar mechanism could exist in plant mitochondria.

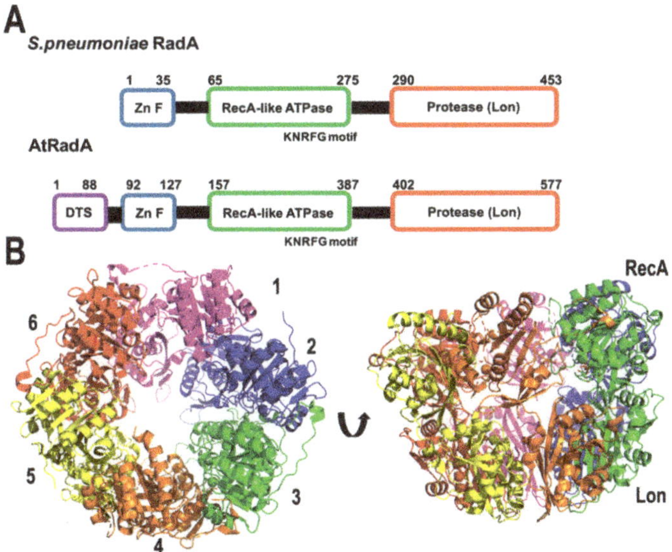

Figure 7. Plant RadA resembles the bacterial enzyme. (**A**) Structural organization of AtRadA in comparision to bacterial RadA. AtRadA shares 63% amino acid similarity with RadA from *S. pneumoniae* and complete amino acid identity in the catalytic amino acids. Bacterial RadA harbor a zinc finger (ZnF), a Rec-A like ATPase domain with a unique KNRFG motif, and a region homologous to the Lon protease. (**B**) Crystal structure of the Rec-A like ATPase and Lon protease domains of RadA from *S. pneumoniae* showing its resemblance to a hexameric helicase. The ZnF domain is not present in the crystal structure.

2.2.5. AtRecG

DNA lesions like thymine-dimers or abasic sites, that potentially block replicative DNA helicases and DNA polymerases, are expected to be predominant in plant mitochondria. Thus, it is expected that plant mitochondria have developed mechanisms to avoid replication roadblocks that lead to replication fork collapse. Stalled replication forks can be resolved via the formation of four-strand Holliday junctions. In bacteria and bacteriophage T4, the helicases RecG and UvsW execute this process [75,116–118]. Bacterial RecGs are loaded in a stalled replication fork where they catalyze replication fork reversal by "pushing" a halted three-strand fork and convert this three-strand fork into a four-strand junction or Holliday junction [117–120]. The Holliday junction structure functions as a starting point for replication fork restart.

Flowering plants encode a RecG homolog that is conserved from green algae [121]. In Arabidopsis this protein consists of 957 amino acids, from those residues its first 57 amino acids correspond to an organellar targeting sequence. AtRecG shares 34% amino acid identity with RecG from *Thermotoga maritima* and is expected to have a similar structure (Figure 8). Arabidopsis plants compromised in their RecG activity are prone to suffer recombination events at intermediate-size repeats and this phenomenon increases in plants deficient in AtRecA3 [121]. Although the precise role of AtRecG is unknown, this protein may be involved in the processing of Holliday junction structures and avoiding replication fork collapse or promoting DNA double-strand break repair.

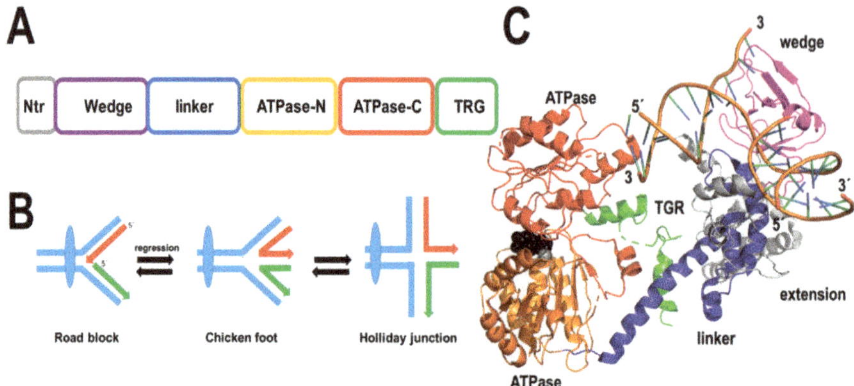

Figure 8. Plants harbor a RecG ortholog. (**A**) AtRecG presents the same domain organization of bacterial RecG, plus the addition of an N-terminal organellar targeting sequence. (**B**) RecG remodels halted replication forks by promoting fork regression (chicken foot structure) that is converted to a Holliday junction. (**C**) Crystal structure of *T. maritima* RecG illustrating its modular assembly.

2.3. Unique Proteins in Flowering Plant Mitochondria

Flowering plant mitochondria have unique proteins. These proteins include: (i) Replicative DNA polymerases solely encoded by protists and plants, (ii) a modified family of single-stranded binding proteins, dubbed organellar single-stranded DNA binding proteins (OSBs) in which their OB-fold suffered extensive modifications, (iii) an associated motif dubbed PDF that plays a role in binding to ssDNA, (iv) a protein that resembles Muts from bacteria, dubbed Msh1, that is only found in plants and corals, and (v) a distinctive family of proteins that belong to a family dubbed whirly (Table 3) [65,122–128]. Both Msh1 and whirlies are proposed to play a dual role in DNA metabolism and as sensor proteins via retrograde signaling from chloroplast-to-nucleus [129,130].

Table 3. Unique proteins involved in DNA metabolism in flowering plant mitochondria.

Arabidopsis Organeles	Acession Number	Localization
AtPolIA	At1g50840	Mitochondria and chloroplasts
AtPolIB	At1g30680	Mitochondria
AtWhy2	At1g71260	Mitochondria
AtOSB1	At1g47720	Mitochondria
AtOSB3	At5g44785	Mitochondria and chloroplasts
AtOSB4	At1g31010	Mitochondria, chloroplasts
AtMsh1	At3g24320.1	Nuclear, mitochondria and chloroplasts

2.3.1. Plant Organellar DNA Polymerases (POPs)

DNA polymerases in metazoan mitochondria are related to bacteriophage T-odd DNA polymerases [12,131]. Pioneering studies by the groups of Professors Sakaguchi and Sato revealed that plant organellar DNA polymerases have a different evolutionary history than phage and mitochondrial DNAPs from metazoans [122–125]. POPs belong to the family A of DNA polymerases; however, they did not evolve from bacteriophage T-odd DNAPs. Flowering plants harbor two paralogous POP genes with chloroplast and mitochondrial localization. In Arabidopsis, one POP is a high-fidelity DNAP (AtPolIA), whereas the other, AtPolIB, is a low-fidelity enzyme [132]. From a structural point of view, the most distinctive elements in POPs are the presence of three unique insertions in their polymerization domain, two of those insertions are located in the thumb subdomain (Ins1 and Ins2), whereas the third insertion is placed in the fingers subdomain [122–125]. Ins1 and Ins3 are involved in lyase, strand-displacement, and MMEJ activities [59,133,134] (Figure 9). AtPolIA and AtPolIB interact with AtTwinkle, and extend primers synthesized by its primase module [29,34]. The physical interaction between AtPolIs with AtTwinkle and AtSSB1 suggests the presence of a functional plant mitochondrial replisome [34]. Biochemical and functional evidence suggests that AtPolIA plays a predominant role in DNA replication, whereas the AtPolIB paralog plays a role in DNA repair [132,135,136]. The gene duplication event in POP evolution suggests a possible event of specialization. This situation resembles the presence of duplicated copies of the replicative DNA polymerase in Mycobacterium, in which one copy contributes to drug resistance because of its low nucleotide incorporation fidelity [137]. In this scenario, AtPolIB could be in the process of becoming a DNAP specialized in translesion synthesis or in other DNA repair pathways. Although AtPolIA and AtPolIB share more than 70% amino acid identity, a single amino acid change in homologous DNA polymerases provides translesion DNA synthesis capabilities [138].

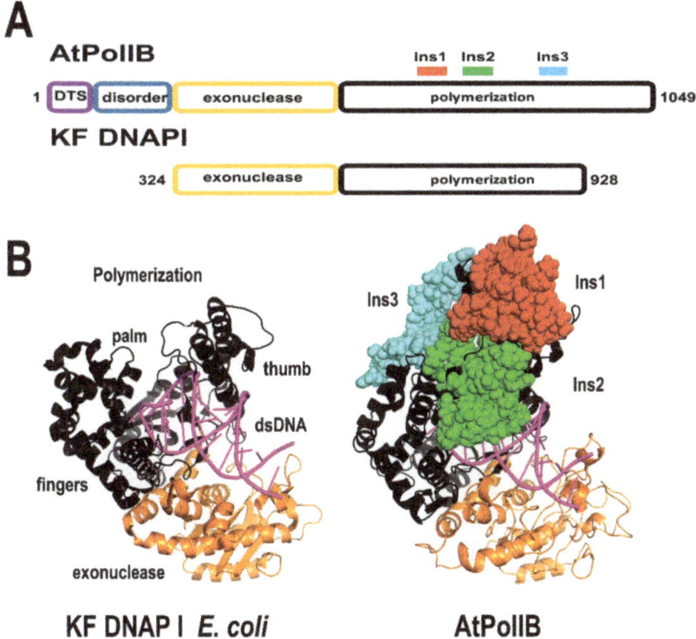

Figure 9. Structural comparison between AtPolIB and bacterial DNAPs. (**A**) Domain organization of both DNAPs. The polymerization domains are colored in black and the 3′-5′ exonuclease domains in orange. The unique amino acid insertions in AtPolIB in comparison to bacterial DNAPs I are depicted in a ball-stick representation and colored in red, green, and cyan. AtPolIs contain an N-terminal DTS and a disorder region not present in the structural model. (**B**) homology model of AtPolIBs with the crystal structures of the Klenow fragment from *E. coli* DNAP I. In both models, the dsDNA from Bacillus DNAP I is superimposed.

2.3.2. AtWhirlies

The most iconic family of single-stranded binding proteins in plant mitochondria is a family dubbed whirly. Whirlies are oligomeric proteins unique to plants. In contrast to the majority of organellar DNA binding proteins, whirlies are encoded in the nucleus and were initially identified as nuclear transcription factors [139]. Whirlies assemble as tetramers, however, upon binding to long-stretches of ssDNA they form a 24-mer assembly [140,141]. Arabidopsis harbors three members of the Whirly family, AtWhy2 localizes to mitochondria, and as a monomer is the most abundant DNA binding protein in plant mitochondria [55], whereas AtWhy1 and AtWhy3 translocate into chloroplasts [55,142]. T-insertional lines of Arabidopsis that knockout AtWhy1 and AtWhy3 accumulate DNA arrangements at microhomologous repeats in the chloroplast [143]. However, Arabidopsis plants devoid of AtWhy2 present a wild-type phenotype and do not accumulate MMEJ products in the absence of agents that induce DSBs [135,144], and show only a small increase in MMEJ products in presence of ciprofloxacin [135].

Whirly proteins bind ssDNA with nanomolar affinity and exhibit a novel protein fold in which each whirly monomer consists of two antiparallel beta sheets organized along two alpha-helices that resembles a whirligig [128,141]. The whirly domain comprises between 150 to 200 amino acids and contains an acidic/aromatic C-terminal end, that is disordered in crystal structures. The residues involved in ssDNA binding are distributed along the two antiparallel beta sheets and whirlies interact with ssDNA via hydrophobic residues and hydrogen bonds mediated by polar amino acids [140] (Figure 10). Whirlies harbor a conserved KGKAAL motif, located in the second beta strand of the first β-sheet, whose integrity is necessary for the 24-mer assembly [140]. Although mutations in this domain do not affect binding to short ssDNA segments, Arabidopsis complemented with a Why construct in which the second lysine of the KGKAAL motif is mutated to alanine are incompetent to reduce the appearance of microhomologies [140]. The latter suggests that the functional oligomeric state of Whirlies in vivo is a 24-mer. The solvent exposed localization of the unstructured C-terminal tail in whirlies suggests that they may mediate protein–protein interactions, analogous to bacterial SSB.

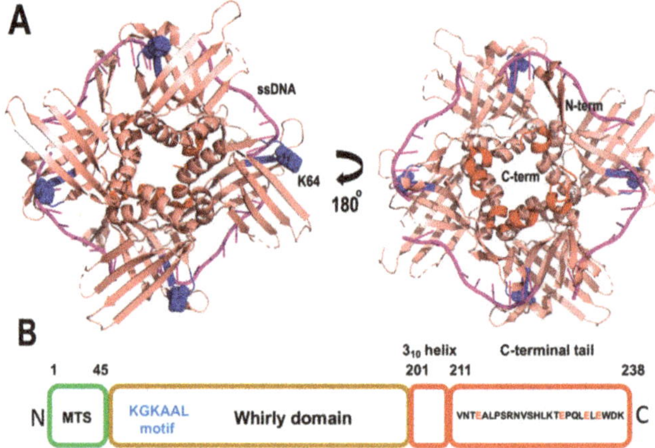

Figure 10. Structural organization of Whirlies. (**A**) Crystal structure of AtWhy2 (PDB ID: 4kop) with model ssDNA from Solanum whirly. The crystal structure represents residues 45 to 212. The second lysine of the KGKAAL motif is in a ball-stick representation. The C-terminal 3_{10} helix is in red. (**B**) Structural organization of AtWhy2. The disordered C-terminal tail is indicated in the diagram.

2.3.3. Organellar Single-Stranded DNA Binding Proteins (OSBs)

The groups of Gualberto and Imbault identified a unique family of single-stranded DNA binding proteins conserved from green algae to flowering plants [126]. These proteins harbor an N-terminal OB-fold domain linked to a motif of 50 amino acids dubbed PDF motif, because of a conserved signature of Pro, Asp, and Phe. Those researchers coined the name "Organellar Single-stranded DNA Binding proteins (OSB)" for members of this protein family. In OSBs, the PDF motif can be arranged as one or multiple copies (Figure 11). Arabidopsis contains four OSBs proteins, dubbed AtOSB1 to AtOSB4. AtOSB1and AtOSB2 are targeted exclusively to mitochondria and chloroplast, respectively, whereas AtOSB3 presents dual-target localization. Quantitative proteomic analysis showed that AtOSB4 and AtOSB3 are highly abundant proteins in mitochondria, whereas AtOSB1 is present at very low concentrations [55]. Remarkably, T-insertion lines of AtOSB1 generate homologous recombination products at repeats that are not commonly used [126].

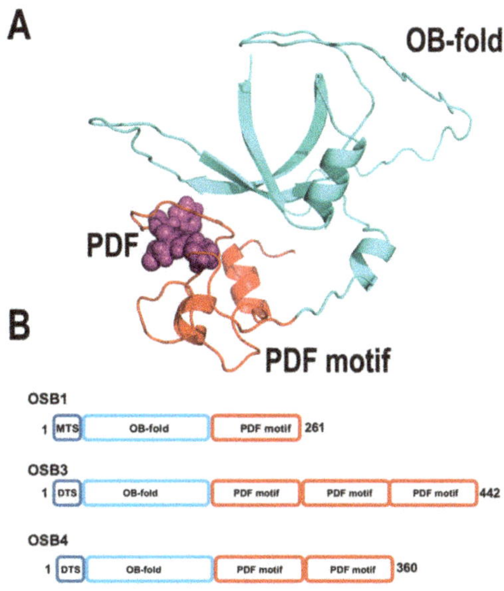

Figure 11. Structural organization of OSBs. (**A**) Structural model of AtOSB1 showing its predicted OB-fold and PDF motif domains. (**B**) Modular organization of mitochondrial OSBs in *Arabidopsis*. AtOSBs consist of an OB-like fold followed by one to three PDF motifs (54). Although AtOSB1 is depicted as a monomer, AtOSB2 in solution assembles as tetramer.

AtOSB2 assembles as a tetramer and binds ssDNA with nanomolar affinity [59]. The PDF motif of AtOSB1 is sufficient for binding to ssDNA, whereas its OB-fold appears to have lost its ability to bind ssDNA [126]. AtOSB2 does not interact with AtPolIs, suggesting that in contrast to other single-stranded binding proteins, its role is not to avoid the formation of secondary structure elements that halt replicative DNA polymerases [59]. The high-affinity of AtOSBs for single-stranded DNA regions and their high abundance within mitochondrial DNA suggest that they coat single-stranded regions of DNA. This coating correlates with the increase of non-canonical homologous recombination products in plants lacking AtOSB1 [126].

2.3.4. AtMhs1

George P. Rédei discovered that the CHLOROPLAST MUTATOR (chm) locus induces plant variegation and impaired fertility, and that both traits are inhered maternally [145,146]. The chm locus regulates the formation of rearrangements in plastids and mitochondria [147] and it encodes for a protein with resemblance to bacterial MutS, and therefore it was named Msh1 [65]. In bacteria, MutS and MutL are conserved elements of the DNA mismatch repair pathway. Within this pathway, MutS recognizes a mismatch and recruits the MutL endonuclease. Recognition of the mismatch correspondingly to the newly synthesized DNA chain is mediated by hemimethylation recognized by MutH [148]. The MSH1 gene is only present in corals and plants and is a multidomain protein harboring domains with homology to bacterial MutS and the GIY-YIG endonuclease [65,127,149,150]. Plants harboring deletions of this gene exhibit increased recombination frequencies at intermediate-size repeats. It is clear that Msh1 guards organellar genomes against aberrant or not frequent recombination events and the roles of Msh1 appear to be related to homeologous recombination suppression [5,68]. Thus, Msh1 resembles a minimal MutS/MutL complex, in which the GIY-YIG endonuclease may play the same role as that MutL endonuclease. In spite of its prevalent role in keeping a pristine plant mitochondrial genome, the only functional study of this protein comes from the characterization of its GIY-YIG domain. By itself this domain binds to branched DNA structures, however the individual domain is not active as an endonuclease [151]. The proposed role of Msh1 in supressing homeologous recombination resembles the role of MutS2 in Helicobacter pylori which harbors an Smr domain that is a non-specific endonuclease [152,153].

2.4. The Bacterial Gyrase, the Eukaryotic DNA Ligase, and the Archaeo-Eukaryotic PrimPol

2.4.1. The Bacterial-Like Plant Organellar Gyrase

Topoisomerases are divided into two types, type I topoisomerases transiently introduce ssDNA breaks and type II transiently generate dsDNA breaks. DNA gyrase is a type II topoisomerase typically present in bacteria. This enzyme is a tetramer encoded by two subunits of the GyrA and GyrB proteins. Bacterial gyrases use ATP to introduce negative supercoils in DNA. Wall and coworkers discovered that flowering plants encode one gene for gyrA (At3g10690) and two functional genes of gyrB (At3g10270 and At5g04130) [154,155]. AtGyrA is targeted to mitochondria and chloroplast, whereas the product of At5g04130 is targeted to mitochondria and was dubbed AtmtGyrB [154]. Both AtGyrA and the two AtmtGyrBs have a clear cyanobacterial origin [154].

Structural studies of bacterial gyrases show the coordination between gyrA and gyrB that drives cleavage of the DNA strands, strand passage between subunits, and ligation [156–158]. Heterologously purified AtGyrA/AtmtGyrB present supercoiling activity [155] and the bacterial origin of the plant organellar AtGyrA/AtmtGyrB makes them a target for the development of new herbicides based on quinolones [155]. Ciprofloxacin, a quinolone drug, is commonly used to induce specific DSBs in plant organelles as the gyrase catalytic cycle is not completed [135,159]. However, bacterial DNA gyrases in complex with quinolone drugs pose a barrier for replication and transcription when bound to DNA and it is possible that the DBS results from the collision of replication forks [160]. As replication induces the formation of positive supercoils ahead of replication forks [161], the plant organellar DNA gyrase may control the formation of origins of replication and the rate of transcription.

2.4.2. Nuclear DNA Ligase I Is Targed to Organelles

Arabidopsis thaliana encodes for three ATP dependent DNA ligases, dubbed DNA ligase I, IV, and VI. From these, DNA ligase I is located in the nucleus and mitochondria. DNA ligase IV is solely nuclear and DNA ligase VI is possibly targeted to both nucleus and chloroplast [162,163]. Thus, in flowering plants, DNA ligase I (At1g08130.1) is the only ligase known to be targeted to mitochondria [163]. DNA ligase I from Arabidopsis (AtDNAligI) shares 46% amino acid identity with DNA ligase I from humans and its mitochondrial targeting sequence is predicted to involve the first 53 amino acids [113]. The unique role of DNA ligase I in plants contrast with the situation in metazoans in which a specific DNA ligase, dubbed DNA ligase III, is the main DNA ligase in human mitochondria. Although this scenario appears to be specific to vertebrates and in lower eukaryotes, DNA ligase I is both a nuclear and a mitochondrial ligase [164,165]. DNA ligases I are structurally divided into three conserved domains: DNA binding, adenylation, and OB-fold. They also contain an N-terminal PCNA interaction motif, as the interaction between DNA ligase I and PCNA is crucial for efficient nick-sealing. Human DNA ligase I have a toroidal shape structure in which PCNA could be accommodated [166].

The ligase active site is assembled between amino acids from the DNA binding and adenylation domains. Those domains harbors six conserved motifs (I, III, IIIa, IV, V, and VI) including the active site lysine, involved in the formation of the ligase–AMP intermediate [166,167]. As flowering plants appear to only have DNA ligase I in their mitochondria, this ligase is predicted to execute all nick sealing reactions. ATLIG1 is an essential gene and besides its role in DNA replication, it is involved in repairing single and DSBs [162]. A homology-based model of *A. thaliana* DNA ligase I using human DNA ligase I shows the predicted fold conservation between both proteins (Figure 12). The PCNA-interacting peptide (PIP box) motif, located at the N-terminal region of DNA ligases, is predicted to be absent in the mitochondrial isoform after its import into mitochondria (Figure 12). Although it is plausible that Arabidopsis DNA ligase I establishes a set of specific protein–protein interactions with protein partners in mitochondria, it is also possible that Arabidopsis DNA ligase I in mitochondria executes nick-sealing without the assistance of accessory proteins. Supporting this scenario, human mitochondrial DNA ligase III can be substituted for bacterial and viral ligases [168].

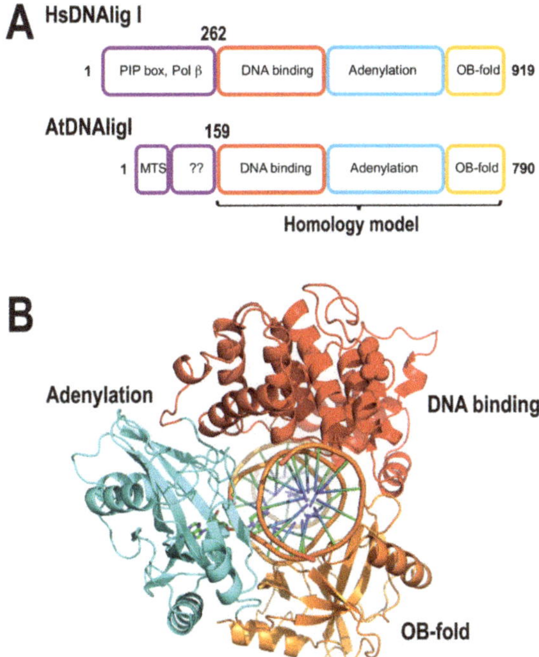

Figure 12. Structural comparison between HsDNAligI and AtDNligI. (**A**) AtDNAligI has a shorter N-terminal region. However, the core structure that harbors the DNA binding domain (red) the adenylation domain (cyan) and the OB-fold domain (orange) are conserved between both ligases. (**B**) Homology modeling of AtDNAlig I with basis on the crystal structure of human DNA ligase I (PDB ID: 1X9N).

2.4.3. Plant PrimPol

Three independent groups discovered that eukaryotic cells harbor a novel primase from the archaeo-eukaryotic primase (AEP) superfamily [169–171]. This enzyme is homologous to eukaryotic primases, but harbors both primase and polymerase activities in a single polypeptide and therefore it was dubbed PrimPol [169–171]. PrimPol contains independent AEP and zinc finger domains; the first domain is responsible for template-dependent nucleotide incorporation and the second domain provides a mechanism to recognize single-stranded DNA templates [170,172–174]. Human PrimPol localizes to the nucleus and mitochondria [170]. In human mitochondria, this enzyme is not involved in primer synthesizes during mitochondrial replication, but in negotiating DNA lesions by repriming and translesion DNA synthesis [169,175]. *Arabidopsis thaliana* harbors a PrimPol ortholog (AtPrimPol -At5g52800-). This enzyme is potentially a translesion synthesis DNA polymerase able of primer synthesis at specific single-stranded DNA sequences (Figure 13). This enzyme harbors localization signal for the nucleus, the mitochondria, and the chloroplast, suggesting that it may play a role in translesion DNA synthesis in each genome.

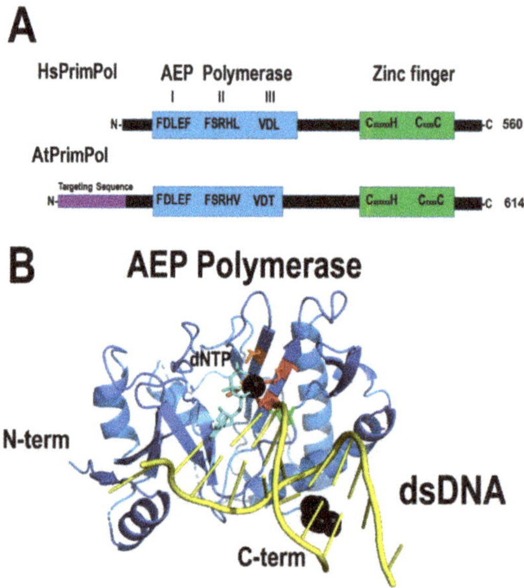

Figure 13. AtPrimPol resembles HsPrimPol. (**A**) Both AtPrimPol and HsPrimPol share a modular organization. AtPrimPol contains an N-terminal sequence for dual organellar targeting. (**B**) Structural model of the archaeo-eukaryotic primase (AEP) domain of AtPrimPol. The structural model was constructed with basis on the crystal structure of the AEP domain of HsPrimPol.

3. Known Unknowns in Plant Mitochondrial Replication

3.1. Mitochondrial DNA Replication Is Mosaic and Redundant

Plant mitochondrial DNA replication is carried out by mosaic and redundant elements (Tables 1–3). For instance, two DNA polymerases (AtPolIA and AtPolIB) are capable of executing DNA replication; at least three different processes may exist for DNA unwinding: (a) Direct unwinding by AtTwinkle, (b) direct unwinding by RadA, and (c) intrinsic unwinding by AtPolIs due to their strong strand-displacement activities; and five different processes (double stranded breaks, abortive transcription by mitochondrial RNA polymerases, and primer synthesis by AtTwinkle, AtTwinky, and AtPrimPol) could generate 3′-OHs needed to start replication. Thus, is not surprising that few genes involved in mitochondrial DNA replication are essential.

In the coordinated leading and lagging-strand DNA synthesis model, an RNA polymerase synthesizes long RNA primers at unknown replication origins, AtTwinkle assembles at the single-stranded region, and these RNA primers are extended by a leading-strand AtPolI. AtTwinkle coordinates leader and lagging-strand synthesis by its primase activity. In the recombination-dependent replication system, a double-stranded break is resected and could be coated with AtRecAs. AtRecA would be responsible to find a homologous region in a double-stranded DNA segment. During AtRecA binding, the plant helicase AtRadA may bind to the single-stranded DNA assembling a replisome upon the interaction with AtPolIA or AtPolB (Figure 14).

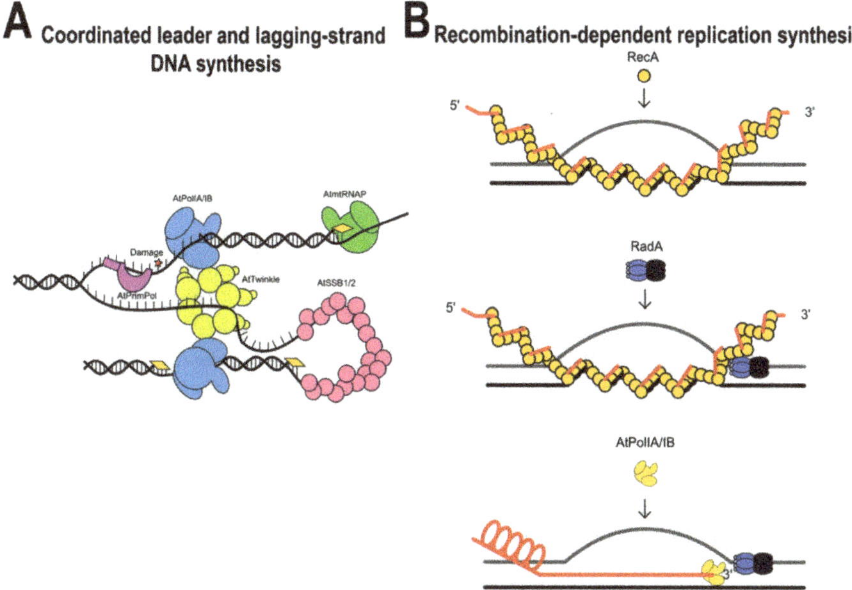

Figure 14. Putative models for DNA replication in plant mitochondria. (**A**) Leader and lagging-strand DNA synthesis. (**B**) Recombination-dependent replication systems in plant mitochondria.

In contrast to metazoan mitochondria, in which the four enzymes responsible for its replication are clearly related to enzymes from T-odd bacteriophages, plant mitochondria harbor enzymes with clear bacterial origin (DNA gyrase), proteins solely present in plant mitochondria (Msh1, OSBs, Why), and enzymes related to bacteriophages (AtTwinkle). This redundant and mosaic system may be responsible for the peculiarities present in plant mitochondrial genomes.

The study of DNA metabolism in plant mitochondria is in its infancy. We do not know how DNA replication in plant mitochondria starts, if plant mitochondria genomes need an origin of replication, and our knowledge of the physical interaction between the proteins involved in mitochondrial DNA metabolism is practically null. The classic view of the need of an origin of replication is given by the study of DNA replication in *E. coli*, where the initiator protein DnaA binds to specific sequences to drive replication initiation. In metazoan mitochondria, its RNA polymerase synthesizes RNA primers that function as primers for heavy and light chains, and it is generally accepted that yeast mitochondria start its replication at double-stranded breaks.

3.2. How Is the Accesibility to Single-Stranded DNA Regulated?

A recent proteomic analysis shows that AtRecA2, AtSSB1, AtSSB2, AtWhy2, AtOSB3, and AtOSB4 are among the most abundant proteins in plant mitochondria [55]. In solution, AtSSB1, AtWhy2, and AtOSB2 assemble as tetramers, although AtOSB2 readily form higher-order complexes (possible 8-mers or 16-mers) [59]. Surprisingly, AtWhy2 assembles as 24-mers in the presence of long segments of ssDNA (more than 7 Kbs) [140]. The carefull study by Fuchs and coworkers reveals that plant mitochondria contains approximately 140 tetramers of AtSSBs, 45 tetramers of AtOSB3 or AtOSB4, and 240 tetramers of AtWhy2 [59,128]. The abundance of AtWhy2 correlates with the fact that plants devoid of this protein accumulate DNA rearrangements mediated by microhomologous regions in the presence of agents that create DSBs [135,140]. Although no cellular studies using AtOSB2 or AtOSB3 have been carried out to date, AtOSB1 mutants accumulate homologous recombination products at repeats that are not commonly used [126]. Given that single-stranded regions of mitochondrial

DNA are coated with AtSSB2s, AtWhy2, AtOSB2, and AtOSB3, it is unknown how these proteins are removed. A possible mechanism involves AtODB1, however AtODB1 lacks the C-terminal domain involved in protein–protein interactions. Thus, it is unknown if AtODB1 is able to displace ssDNA binding proteins like AtWhy2, AtSSBs, or AtOSBs from ssDNA or if AtSSBs interact with AtRecA2 to promote filament assembly.

3.3. Open Question in Plant Mitochondrial DNA Replication

It is puzzling how the open reading frames in plant mitochondria exhibit low substitution rates, while their non-coding regions are highly variable [6,9]. Mitochondrial DNA in land plants exists as linear molecules and it is proposed that neighboring DNA molecules can act as a template to avoid mutations [6]. If this is the case, it is unknown how the correct sequence is selected, given that plant mitochondrial DNA is not methylated. Furthermore, plant organellar DNA polymerases in Arabidopsis present a gradient of almost 10-fold in replication fidelity [134] and it is unknown if postraslational modification can affect their interaction with other proteins and their biochemical properties.

Several studies indicate the presence of non-homologous end joining (NHEJ) repair signatures in plant mitochondria. However, the key components of this route Artemis and Ku proteins are not targeted to plant mitochondria and the mechanisms by which a NHEJ-like route operate in plant mitochondria are unknown. Recent work using hybrid mitochondrial cell lines discovered that changes in the human epigenome are driven by modifications in the mitochondrial genome [176]. Does the highly recombinogenic nature of plant mitochondrial DNA confers an evolutionary advantage for flowering plants as a hub for adaptation?

Funding: Work in L.G.B. laboratory is supported by grants SEP-CINVESTAV-63 and CONACYT-253737.

Acknowledgments: Cei Abreu for critical reading and Víctor Juárez for Figure 14.

Conflicts of Interest: The author declares no conflict of interest.

References

1. Gray, M.W. Mitochondrial evolution. *Cold Spring Harb. Perspect. Biol.* **2012**, *4*, a011403. [CrossRef] [PubMed]
2. Smith, D.R.; Keeling, P.J. Mitochondrial and plastid genome architecture: Reoccurring themes, but significant differences at the extremes. *Proc. Natl. Acad. Sci. USA* **2015**, *112*, 10177–10184. [CrossRef] [PubMed]
3. Gissi, C.; Iannelli, F.; Pesole, G. Evolution of the mitochondrial genome of Metazoa as exemplified by comparison of congeneric species. *Heredity* **2008**, *101*, 301–320. [CrossRef] [PubMed]
4. Boore, J.L. Animal mitochondrial genomes. *Nucleic Acids Res.* **1999**, *27*, 1767–1780. [CrossRef] [PubMed]
5. Davila, J.I.; Arrieta-Montiel, M.P.; Wamboldt, Y.; Cao, J.; Hagmann, J.; Shedge, V.; Xu, Y.Z.; Weigel, D.; Mackenzie, S.A. Double-strand break repair processes drive evolution of the mitochondrial genome in Arabidopsis. *BMC Biol.* **2011**, *9*, 64. [CrossRef]
6. Christensen, A.C. Genes and junk in plant mitochondria-repair mechanisms and selection. *Genome Biol. Evol.* **2014**, *6*, 1448–1453. [CrossRef]
7. Sloan, D.B.; Wu, Z.; Sharbrough, J. Correction of Persistent Errors in Arabidopsis Reference Mitochondrial Genomes. *Plant Cell* **2018**, *30*, 525–527. [CrossRef]
8. Unseld, M.; Marienfeld, J.R.; Brandt, P.; Brennicke, A. The mitochondrial genome of Arabidopsis thaliana contains 57 genes in 366,924 nucleotides. *Nat. Genet.* **1997**, *15*, 57–61. [CrossRef]
9. Christensen, A.C. Plant mitochondrial genome evolution can be explained by DNA repair mechanisms. *Genome Biol. Evol.* **2013**, *5*, 1079–1086. [CrossRef]
10. Ruiz-Pesini, E.; Lott, M.T.; Procaccio, V.; Poole, J.C.; Brandon, M.C.; Mishmar, D.; Yi, C.; Kreuziger, J.; Baldi, P.; Wallace, D.C. An enhanced MITOMAP with a global mtDNA mutational phylogeny. *Nucleic Acids Res.* **2007**, *35*, D823–D828. [CrossRef]
11. Chinnery, P.F.; Hudson, G. Mitochondrial genetics. *Br. Med. Bull.* **2013**, *106*, 135–159. [CrossRef] [PubMed]
12. Lang, B.F.; Gray, M.W.; Burger, G. Mitochondrial genome evolution and the origin of eukaryotes. *Annu. Rev. Genet.* **1999**, *33*, 351–397. [CrossRef] [PubMed]

13. Shutt, T.E.; Gray, M.W. Bacteriophage origins of mitochondrial replication and transcription proteins. *Trends Genet.* **2006**, *22*, 90–95. [CrossRef] [PubMed]
14. Xu, B.; Clayton, D.A. RNA-DNA hybrid formation at the human mitochondrial heavy-strand origin ceases at replication start sites: An implication for RNA-DNA hybrids serving as primers. *EMBO J.* **1996**, *15*, 3135–3143. [CrossRef]
15. Fuste, J.M.; Wanrooij, S.; Jemt, E.; Granycome, C.E.; Cluett, T.J.; Shi, Y.; Atanassova, N.; Holt, I.J.; Gustafsson, C.M.; Falkenberg, M. Mitochondrial RNA polymerase is needed for activation of the origin of light-strand DNA replication. *Mol. Cell* **2010**, *37*, 67–78. [CrossRef]
16. Yasukawa, T.; Reyes, A.; Cluett, T.J.; Yang, M.Y.; Bowmaker, M.; Jacobs, H.T.; Holt, I.J. Replication of vertebrate mitochondrial DNA entails transient ribonucleotide incorporation throughout the lagging strand. *EMBO J.* **2006**, *25*, 5358–5371. [CrossRef]
17. Reyes, A.; Kazak, L.; Wood, S.R.; Yasukawa, T.; Jacobs, H.T.; Holt, I.J. Mitochondrial DNA replication proceeds via a 'bootlace' mechanism involving the incorporation of processed transcripts. *Nucleic Acids Res.* **2013**, *41*, 5837–5850. [CrossRef]
18. Miralles Fuste, J.; Shi, Y.; Wanrooij, S.; Zhu, X.; Jemt, E.; Persson, O.; Sabouri, N.; Gustafsson, C.M.; Falkenberg, M. In vivo occupancy of mitochondrial single-stranded DNA binding protein supports the strand displacement mode of DNA replication. *PLoS Genet.* **2014**, *10*, e1004832. [CrossRef]
19. Holt, I.J.; Lorimer, H.E.; Jacobs, H.T. Coupled leading- and lagging-strand synthesis of mammalian mitochondrial DNA. *Cell* **2000**, *100*, 515–524. [CrossRef]
20. Bowmaker, M.; Yang, M.Y.; Yasukawa, T.; Reyes, A.; Jacobs, H.T.; Huberman, J.A.; Holt, I.J. Mammalian mitochondrial DNA replicates bidirectionally from an initiation zone. *J. Biol. Chem.* **2003**, *278*, 50961–50969. [CrossRef]
21. Cluett, T.J.; Akman, G.; Reyes, A.; Kazak, L.; Mitchell, A.; Wood, S.R.; Spinazzola, A.; Spelbrink, J.N.; Holt, I.J. Transcript availability dictates the balance between strand-asynchronous and strand-coupled mitochondrial DNA replication. *Nucleic Acids Res.* **2018**, *46*, 10771–10781. [CrossRef] [PubMed]
22. Lee, S.J.; Richardson, C.C. Choreography of bacteriophage T7 DNA replication. *Curr. Opin. Chem. Biol.* **2011**, *15*, 580–586. [CrossRef] [PubMed]
23. Falkenberg, M. Mitochondrial DNA replication in mammalian cells: overview of the pathway. *Essays Biochem.* **2018**, *62*, 287–296. [CrossRef] [PubMed]
24. Spelbrink, J.N.; Li, F.Y.; Tiranti, V.; Nikali, K.; Yuan, Q.P.; Tariq, M.; Wanrooij, S.; Garrido, N.; Comi, G.; Morandi, L.; et al. Human mitochondrial DNA deletions associated with mutations in the gene encoding Twinkle, a phage T7 gene 4-like protein localized in mitochondria. *Nat. Genet.* **2001**, *28*, 223–231. [CrossRef]
25. Hamdan, S.M.; Loparo, J.J.; Takahashi, M.; Richardson, C.C.; van Oijen, A.M. Dynamics of DNA replication loops reveal temporal control of lagging-strand synthesis. *Nature* **2009**, *457*, 336–339. [CrossRef]
26. Hamdan, S.M.; Richardson, C.C. Motors, switches, and contacts in the replisome. *Annu. Rev. Biochem.* **2009**, *78*, 205–243. [CrossRef]
27. O'Donnell, M.; Langston, L.; Stillman, B. Principles and concepts of DNA replication in bacteria, archaea, and eukarya. *Cold Spring Harb. Perspect. Biol.* **2013**, *5*. [CrossRef]
28. Diray-Arce, J.; Liu, B.; Cupp, J.D.; Hunt, T.; Nielsen, B.L. The Arabidopsis At1g30680 gene encodes a homologue to the phage T7 gp4 protein that has both DNA primase and DNA helicase activities. *BMC Plant. Biol.* **2013**, *13*, 36. [CrossRef]
29. Peralta-Castro, A.; Baruch-Torres, N.; Brieba, L.G. Plant organellar DNA primase-helicase synthesizes RNA primers for organellar DNA polymerases using a unique recognition sequence. *Nucleic Acids Res.* **2017**, *45*, 10764–10774. [CrossRef]
30. Shutt, T.E.; Gray, M.W. Twinkle, the mitochondrial replicative DNA helicase, is widespread in the eukaryotic radiation and may also be the mitochondrial DNA primase in most eukaryotes. *J. Mol. Evol.* **2006**, *62*, 588–599. [CrossRef]
31. Toth, E.A.; Li, Y.; Sawaya, M.R.; Cheng, Y.; Ellenberger, T. The crystal structure of the bifunctional primase-helicase of bacteriophage T7. *Mol. Cell* **2003**, *12*, 1113–1123. [CrossRef]
32. Sawaya, M.R.; Guo, S.; Tabor, S.; Richardson, C.C.; Ellenberger, T. Crystal structure of the helicase domain from the replicative helicase-primase of bacteriophage T7. *Cell* **1999**, *99*, 167–177. [CrossRef]
33. Ilyina, T.V.; Gorbalenya, A.E.; Koonin, E.V. Organization and evolution of bacterial and bacteriophage primase-helicase systems. *J. Mol. Evol.* **1992**, *34*, 351–357. [CrossRef] [PubMed]

34. Morley, S.A.; Peralta-Castro, A.; Brieba, L.G.; Miller, J.; Ong, K.L.; Ridge, P.G.; Oliphant, A.; Aldous, S.; Nielsen, B.L. Arabidopsis thaliana organelles mimic the T7 phage DNA replisome with specific interactions between Twinkle protein and DNA polymerases Pol1A and Pol1B. *BMC Plant. Biol.* **2019**, *19*, 241. [CrossRef] [PubMed]

35. Masters, B.S.; Stohl, L.L.; Clayton, D.A. Yeast mitochondrial RNA polymerase is homologous to those encoded by bacteriophages T3 and T7. *Cell* **1987**, *51*, 89–99. [CrossRef]

36. Kuhn, K.; Richter, U.; Meyer, E.H.; Delannoy, E.; de Longevialle, A.F.; O'Toole, N.; Borner, T.; Millar, A.H.; Small, I.D.; Whelan, J. Phage-type RNA polymerase RPOTmp performs gene-specific transcription in mitochondria of Arabidopsis thaliana. *Plant Cell* **2009**, *21*, 2762–2779. [CrossRef]

37. Hedtke, B.; Borner, T.; Weihe, A. Mitochondrial and chloroplast phage-type RNA polymerases in Arabidopsis. *Science* **1997**, *277*, 809–811. [CrossRef]

38. Bohne, A.V.; Teubner, M.; Liere, K.; Weihe, A.; Borner, T. In vitro promoter recognition by the catalytic subunit of plant phage-type RNA polymerases. *Plant Mol. Biol.* **2016**, *92*, 357–369. [CrossRef]

39. Matsunaga, M.; Jaehning, J.A. Intrinsic promoter recognition by a "core" RNA polymerase. *J. Biol. Chem.* **2004**, *279*, 44239–44242. [CrossRef]

40. Hillen, H.S.; Morozov, Y.I.; Sarfallah, A.; Temiakov, D.; Cramer, P. Structural Basis of Mitochondrial Transcription Initiation. *Cell* **2017**, *171*, 1072–1081.e1010. [CrossRef]

41. Fuller, C.W.; Richardson, C.C. Initiation of DNA replication at the primary origin of bacteriophage T7 by purified proteins. Initiation of bidirectional synthesis. *J. Biol. Chem.* **1985**, *260*, 3197–3206. [PubMed]

42. Fuller, C.W.; Richardson, C.C. Initiation of DNA replication at the primary origin of bacteriophage T7 by purified proteins. Site and direction of initial DNA synthesis. *J. Biol. Chem.* **1985**, *260*, 3185–3196. [PubMed]

43. Wanrooij, S.; Fuste, J.M.; Farge, G.; Shi, Y.; Gustafsson, C.M.; Falkenberg, M. Human mitochondrial RNA polymerase primes lagging-strand DNA synthesis in vitro. *Proc. Natl. Acad. Sci. USA* **2008**, *105*, 11122–11127. [CrossRef] [PubMed]

44. Arimura, S.I. Fission and Fusion of Plant Mitochondria, and Genome Maintenance. *Plant Physiol.* **2018**, *176*, 152–161. [CrossRef]

45. Oldenburg, D.J.; Bendich, A.J. Size and Structure of Replicating Mitochondrial DNA in Cultured Tobacco Cells. *Plant Cell* **1996**, *8*, 447–461. [CrossRef]

46. Backert, S.; Borner, T. Phage T4-like intermediates of DNA replication and recombination in the mitochondria of the higher plant Chenopodium album (L.). *Curr. Genet.* **2000**, *37*, 304–314. [CrossRef]

47. Yang, Z.; Hou, Q.; Cheng, L.; Xu, W.; Hong, Y.; Li, S.; Sun, Q. RNase H1 Cooperates with DNA Gyrases to Restrict R-Loops and Maintain Genome Integrity in Arabidopsis Chloroplasts. *Plant Cell* **2017**, *29*, 2478–2497. [CrossRef]

48. Wu, C.C.; Lin, J.L.J.; Yang-Yen, H.F.; Yuan, H.S. A unique exonuclease ExoG cleaves between RNA and DNA in mitochondrial DNA replication. *Nucleic Acids Res.* **2019**, *47*, 5405–5419. [CrossRef]

49. Liu, P.; Qian, L.; Sung, J.S.; de Souza-Pinto, N.C.; Zheng, L.; Bogenhagen, D.F.; Bohr, V.A.; Wilson, D.M., 3rd; Shen, B.; Demple, B. Removal of oxidative DNA damage via FEN1-dependent long-patch base excision repair in human cell mitochondria. *Mol. Cell Biol.* **2008**, *28*, 4975–4987. [CrossRef]

50. Zheng, L.; Zhou, M.; Guo, Z.; Lu, H.; Qian, L.; Dai, H.; Qiu, J.; Yakubovskaya, E.; Bogenhagen, D.F.; Demple, B.; et al. Human DNA2 is a mitochondrial nuclease/helicase for efficient processing of DNA replication and repair intermediates. *Mol. Cell* **2008**, *32*, 325–336. [CrossRef]

51. Al-Behadili, A.; Uhler, J.P.; Berglund, A.K.; Peter, B.; Doimo, M.; Reyes, A.; Wanrooij, S.; Zeviani, M.; Falkenberg, M. A two-nuclease pathway involving RNase H1 is required for primer removal at human mitochondrial OriL. *Nucleic Acids Res.* **2018**, *46*, 9471–9483. [CrossRef] [PubMed]

52. Uhler, J.P.; Thorn, C.; Nicholls, T.J.; Matic, S.; Milenkovic, D.; Gustafsson, C.M.; Falkenberg, M. MGME1 processes flaps into ligatable nicks in concert with DNA polymerase gamma during mtDNA replication. *Nucleic Acids Res.* **2016**, *44*, 5861–5871. [CrossRef] [PubMed]

53. Cheetham, G.M.; Steitz, T.A. Structure of a transcribing T7 RNA polymerase initiation complex. *Science* **1999**, *286*, 2305–2309. [CrossRef] [PubMed]

54. Edmondson, A.C.; Song, D.; Alvarez, L.A.; Wall, M.K.; Almond, D.; McClellan, D.A.; Maxwell, A.; Nielsen, B.L. Characterization of a mitochondrially targeted single-stranded DNA-binding protein in Arabidopsis thaliana. *Mol. Genet. Genom.* **2005**, *273*, 115–122. [CrossRef] [PubMed]

55. Fuchs, P.; Rugen, N.; Carrie, C.; Elsässer, M.; Finkemeier, I.; Giese, J.; Hildebrandt, T.M.; Kühn, K.; Maurino, V.G.; Ruberti, C.; et al. Single organelle function and organization as estimated from Arabidopsis mitochondrial proteomics. *Plant J.* **2019**, in press. [CrossRef] [PubMed]

56. Khamis, M.I.; Casas-Finet, J.R.; Maki, A.H.; Murphy, J.B.; Chase, J.W. Role of tryptophan 54 in the binding of E. coli single-stranded DNA-binding protein to single-stranded polynucleotides. *FEBS Lett.* **1987**, *211*, 155–159. [CrossRef]

57. Casas-Finet, J.R.; Khamis, M.I.; Maki, A.H.; Chase, J.W. Tryptophan 54 and phenylalanine 60 are involved synergistically in the binding of E. coli SSB protein to single-stranded polynucleotides. *FEBS Lett.* **1987**, *220*, 347–352. [CrossRef]

58. Raghunathan, S.; Kozlov, A.G.; Lohman, T.M.; Waksman, G. Structure of the DNA binding domain of E. coli SSB bound to ssDNA. *Nat. Struct. Biol.* **2000**, *7*, 648–652. [CrossRef]

59. Garcia-Medel, P.L.; Baruch-Torres, N.; Peralta-Castro, A.; Trasvina-Arenas, C.H.; Torres-Larios, A.; Brieba, L.G. Plant organellar DNA polymerases repair double-stranded breaks by microhomology-mediated end-joining. *Nucleic Acids Res.* **2019**, *47*, 3028–3044. [CrossRef]

60. Shinn, M.K.; Kozlov, A.G.; Nguyen, B.; Bujalowski, W.M.; Lohman, T.M. Are the intrinsically disordered linkers involved in SSB binding to accessory proteins? *Nucleic Acids Res.* **2019**. [CrossRef]

61. Kozlov, A.G.; Jezewska, M.J.; Bujalowski, W.; Lohman, T.M. Binding specificity of Escherichia coli single-stranded DNA binding protein for the chi subunit of DNA pol III holoenzyme and PriA helicase. *Biochemistry* **2010**, *49*, 3555–3566. [CrossRef] [PubMed]

62. Stern, D.B.; Palmer, J.D. Recombination sequences in plant mitochondrial genomes: diversity and homologies to known mitochondrial genes. *Nucleic Acids Res.* **1984**, *12*, 6141–6157. [CrossRef] [PubMed]

63. Palmer, J.D.; Adams, K.L.; Cho, Y.; Parkinson, C.L.; Qiu, Y.L.; Song, K. Dynamic evolution of plant mitochondrial genomes: Mobile genes and introns and highly variable mutation rates. *Proc. Natl. Acad. Sci. USA* **2000**, *97*, 6960–6966. [CrossRef] [PubMed]

64. Gualberto, J.M.; Newton, K.J. Plant Mitochondrial Genomes: Dynamics and Mechanisms of Mutation. *Annu. Rev. Plant. Biol.* **2017**, *68*, 225–252. [CrossRef]

65. Abdelnoor, R.V.; Yule, R.; Elo, A.; Christensen, A.C.; Meyer-Gauen, G.; Mackenzie, S.A. Substoichiometric shifting in the plant mitochondrial genome is influenced by a gene homologous to MutS. *Proc. Natl. Acad. Sci. USA* **2003**, *100*, 5968–5973. [CrossRef]

66. Palmer, J.D.; Herbon, L.A. Tricircular mitochondrial genomes of Brassica and Raphanus: Reversal of repeat configurations by inversion. *Nucleic Acids Res.* **1986**, *14*, 9755–9764. [CrossRef]

67. Lonsdale, D.M.; Hodge, T.P.; Fauron, C.M. The physical map and organisation of the mitochondrial genome from the fertile cytoplasm of maize. *Nucleic Acids Res.* **1984**, *12*, 9249–9261. [CrossRef]

68. Arrieta-Montiel, M.P.; Shedge, V.; Davila, J.; Christensen, A.C.; Mackenzie, S.A. Diversity of the Arabidopsis mitochondrial genome occurs via nuclear-controlled recombination activity. *Genetics* **2009**, *183*, 1261–1268. [CrossRef]

69. Shedge, V.; Arrieta-Montiel, M.; Christensen, A.C.; Mackenzie, S.A. Plant mitochondrial recombination surveillance requires unusual RecA and MutS homologs. *Plant Cell* **2007**, *19*, 1251–1264. [CrossRef]

70. Chen, J.; Guan, R.; Chang, S.; Du, T.; Zhang, H.; Xing, H. Substoichiometrically different mitotypes coexist in mitochondrial genomes of Brassica napus L. *PLoS ONE* **2011**, *6*, e17662. [CrossRef]

71. Tang, H.; Zheng, X.; Li, C.; Xie, X.; Chen, Y.; Chen, L.; Zhao, X.; Zheng, H.; Zhou, J.; Ye, S.; et al. Multi-step formation, evolution, and functionalization of new cytoplasmic male sterility genes in the plant mitochondrial genomes. *Cell Res.* **2017**, *27*, 130–146. [CrossRef] [PubMed]

72. Janska, H.; Sarria, R.; Woloszynska, M.; Arrieta-Montiel, M.; Mackenzie, S.A. Stoichiometric shifts in the common bean mitochondrial genome leading to male sterility and spontaneous reversion to fertility. *Plant Cell* **1998**, *10*, 1163–1180. [CrossRef] [PubMed]

73. Kozik, A.; Rowan, B.A.; Lavelle, D.; Berke, L.; Schranz, M.E.; Michelmore, R.W.; Christensen, A.C. The alternative reality of plant mitochondrial DNA: One ring does not rule them all. *PLoS Genet.* **2019**, *15*, e1008373. [CrossRef] [PubMed]

74. Cole, L.W.; Guo, W.; Mower, J.P.; Palmer, J.D. High and Variable Rates of Repeat-Mediated Mitochondrial Genome Rearrangement in a Genus of Plants. *Mol. Biol. Evol.* **2018**, *35*, 2773–2785. [CrossRef] [PubMed]

75. Liu, J.; Morrical, S.W. Assembly and dynamics of the bacteriophage T4 homologous recombination machinery. *Virol. J.* **2010**, *7*, 357. [CrossRef] [PubMed]

76. Bendich, A.J. Structural analysis of mitochondrial DNA molecules from fungi and plants using moving pictures and pulsed-field gel electrophoresis. *J. Mol. Biol.* **1996**, *255*, 564–588. [CrossRef]

77. Cheng, N.; Lo, Y.S.; Ansari, M.I.; Ho, K.C.; Jeng, S.T.; Lin, N.S.; Dai, H. Correlation between mtDNA complexity and mtDNA replication mode in developing cotyledon mitochondria during mung bean seed germination. *New Phytol.* **2017**, *213*, 751–763. [CrossRef]

78. Cox, M.M. Regulation of bacterial RecA protein function. *Crit. Rev. Biochem. Mol. Biol.* **2007**, *42*, 41–63. [CrossRef]

79. Lee, J.Y.; Terakawa, T.; Qi, Z.; Steinfeld, J.B.; Redding, S.; Kwon, Y.; Gaines, W.A.; Zhao, W.; Sung, P.; Greene, E.C. DNA RECOMBINATION. Base triplet stepping by the Rad51/RecA family of recombinases. *Science* **2015**, *349*, 977–981. [CrossRef]

80. Hsieh, P.; Camerini-Otero, C.S.; Camerini-Otero, R.D. The synapsis event in the homologous pairing of DNAs: RecA recognizes and pairs less than one helical repeat of DNA. *Proc. Natl. Acad. Sci. USA* **1992**, *89*, 6492–6496. [CrossRef]

81. Ragunathan, K.; Liu, C.; Ha, T. RecA filament sliding on DNA facilitates homology search. *Elife* **2012**, *1*, e00067. [CrossRef] [PubMed]

82. Story, R.M.; Weber, I.T.; Steitz, T.A. The structure of the *E. coli* recA protein monomer and polymer. *Nature* **1992**, *355*, 318–325. [CrossRef] [PubMed]

83. Chen, Z.; Yang, H.; Pavletich, N.P. Mechanism of homologous recombination from the RecA-ssDNA/dsDNA structures. *Nature* **2008**, *453*, 489–494. [CrossRef] [PubMed]

84. Lin, Z.; Kong, H.; Nei, M.; Ma, H. Origins and evolution of the recA/RAD51 gene family: Evidence for ancient gene duplication and endosymbiotic gene transfer. *Proc. Natl. Acad. Sci. USA* **2006**, *103*, 10328–10333. [CrossRef]

85. Khazi, F.R.; Edmondson, A.C.; Nielsen, B.L. An Arabidopsis homologue of bacterial RecA that complements an *E. Coli* recA deletion is targeted to plant mitochondria. *Mol. Genet. Genom.* **2003**, *269*, 454–463. [CrossRef]

86. Miller-Messmer, M.; Kuhn, K.; Bichara, M.; Le Ret, M.; Imbault, P.; Gualberto, J.M. RecA-dependent DNA repair results in increased heteroplasmy of the Arabidopsis mitochondrial genome. *Plant Physiol.* **2012**, *159*, 211–226. [CrossRef]

87. Odahara, M.; Kuroiwa, H.; Kuroiwa, T.; Sekine, Y. Suppression of repeat-mediated gross mitochondrial genome rearrangements by RecA in the moss Physcomitrella patens. *Plant Cell* **2009**, *21*, 1182–1194. [CrossRef]

88. Rowan, B.A.; Oldenburg, D.J.; Bendich, A.J. RecA maintains the integrity of chloroplast DNA molecules in Arabidopsis. *J. Exp. Bot.* **2010**, *61*, 2575–2588. [CrossRef]

89. Eggler, A.L.; Lusetti, S.L.; Cox, M.M. The C terminus of the Escherichia coli RecA protein modulates the DNA binding competition with single-stranded DNA-binding protein. *J. Biol. Chem.* **2003**, *278*, 16389–16396. [CrossRef]

90. Ithurbide, S.; Bentchikou, E.; Coste, G.; Bost, B.; Servant, P.; Sommer, S. Single Strand Annealing Plays a Major Role in RecA-Independent Recombination between Repeated Sequences in the Radioresistant Deinococcus radiodurans Bacterium. *PLoS Genet.* **2015**, *11*, e1005636. [CrossRef]

91. De Mot, R.; Schoofs, G.; Vanderleyden, J. A putative regulatory gene downstream of recA is conserved in gram-negative and gram-positive bacteria. *Nucleic Acids Res.* **1994**, *22*, 1313–1314. [CrossRef] [PubMed]

92. VanLoock, M.S.; Yu, X.; Yang, S.; Galkin, V.E.; Huang, H.; Rajan, S.S.; Anderson, W.F.; Stohl, E.A.; Seifert, H.S.; Egelman, E.H. Complexes of RecA with LexA and RecX differentiate between active and inactive RecA nucleoprotein filaments. *J. Mol. Biol.* **2003**, *333*, 345–354. [CrossRef] [PubMed]

93. Ragone, S.; Maman, J.D.; Furnham, N.; Pellegrini, L. Structural basis for inhibition of homologous recombination by the RecX protein. *EMBO J.* **2008**, *27*, 2259–2269. [CrossRef] [PubMed]

94. Yang, C.Y.; Chin, K.H.; Yang, M.T.; Wang, A.H.; Chou, S.H. Crystal structure of RecX: A potent regulatory protein of RecA from Xanthomonas campestris. *Proteins* **2009**, *74*, 530–537. [CrossRef]

95. Drees, J.C.; Lusetti, S.L.; Chitteni-Pattu, S.; Inman, R.B.; Cox, M.M. A RecA filament capping mechanism for RecX protein. *Mol. Cell* **2004**, *15*, 789–798. [CrossRef]

96. Cardenas, P.P.; Carrasco, B.; Defeu Soufo, C.; Cesar, C.E.; Herr, K.; Kaufenstein, M.; Graumann, P.L.; Alonso, J.C. RecX facilitates homologous recombination by modulating RecA activities. *PLoS Genet.* **2012**, *8*, e1003126. [CrossRef]

97. Odahara, M.; Sekine, Y. RECX Interacts with Mitochondrial RECA to Maintain Mitochondrial Genome Stability. *Plant Physiol.* **2018**, *177*, 300–310. [CrossRef]

98. Bell, J.C.; Kowalczykowski, S.C. RecA: Regulation and Mechanism of a Molecular Search Engine. *Trends Biochem. Sci.* **2016**, *41*, 491–507. [CrossRef]

99. Vermel, M.; Guermann, B.; Delage, L.; Grienenberger, J.M.; Marechal-Drouard, L.; Gualberto, J.M. A family of RRM-type RNA-binding proteins specific to plant mitochondria. *Proc. Natl. Acad. Sci. USA* **2002**, *99*, 5866–5871. [CrossRef]

100. Janicka, S.; Kuhn, K.; Le Ret, M.; Bonnard, G.; Imbault, P.; Augustyniak, H.; Gualberto, J.M. A RAD52-like single-stranded DNA binding protein affects mitochondrial DNA repair by recombination. *Plant J.* **2012**, *72*, 423–435. [CrossRef]

101. Rendekova, J.; Ward, T.A.; Simonicova, L.; Thomas, P.H.; Nosek, J.; Tomaska, L.; McHugh, P.J.; Chovanec, M. Mgm101: A double-duty Rad52-like protein. *Cell Cycle* **2016**, *15*, 3169–3176. [CrossRef] [PubMed]

102. Pevala, V.; Truban, D.; Bauer, J.A.; Kostan, J.; Kunova, N.; Bellova, J.; Brandstetter, M.; Marini, V.; Krejci, L.; Tomaska, L.; et al. The structure and DNA-binding properties of Mgm101 from a yeast with a linear mitochondrial genome. *Nucleic Acids Res.* **2016**, *44*, 2227–2239. [CrossRef] [PubMed]

103. Hanamshet, K.; Mazina, O.M.; Mazin, A.V. Reappearance from Obscurity: Mammalian Rad52 in Homologous Recombination. *Genes* **2016**, *7*, 63. [CrossRef] [PubMed]

104. Kagawa, W.; Kurumizaka, H.; Ishitani, R.; Fukai, S.; Nureki, O.; Shibata, T.; Yokoyama, S. Crystal structure of the homologous-pairing domain from the human Rad52 recombinase in the undecameric form. *Mol. Cell* **2002**, *10*, 359–371. [CrossRef]

105. Stasiak, A.Z.; Larquet, E.; Stasiak, A.; Muller, S.; Engel, A.; Van Dyck, E.; West, S.C.; Egelman, E.H. The human Rad52 protein exists as a heptameric ring. *Curr. Biol.* **2000**, *10*, 337–340. [CrossRef]

106. Saotome, M.; Saito, K.; Yasuda, T.; Ohtomo, H.; Sugiyama, S.; Nishimura, Y.; Kurumizaka, H.; Kagawa, W. Structural Basis of Homology-Directed DNA Repair Mediated by RAD52. *iScience* **2018**, *3*, 50–62. [CrossRef]

107. Song, B.; Sung, P. Functional interactions among yeast Rad51 recombinase, Rad52 mediator, and replication protein A in DNA strand exchange. *J. Biol. Chem.* **2000**, *275*, 15895–15904. [CrossRef]

108. Sugiyama, T.; Kowalczykowski, S.C. Rad52 protein associates with replication protein A (RPA)-single-stranded DNA to accelerate Rad51-mediated displacement of RPA and presynaptic complex formation. *J. Biol. Chem.* **2002**, *277*, 31663–31672. [CrossRef]

109. Bhargava, R.; Onyango, D.O.; Stark, J.M. Regulation of Single-Strand Annealing and its Role in Genome Maintenance. *Trends Genet.* **2016**, *32*, 566–575. [CrossRef]

110. Grimme, J.M.; Honda, M.; Wright, R.; Okuno, Y.; Rothenberg, E.; Mazin, A.V.; Ha, T.; Spies, M. Human Rad52 binds and wraps single-stranded DNA and mediates annealing via two hRad52-ssDNA complexes. *Nucleic Acids Res.* **2010**, *38*, 2917–2930. [CrossRef]

111. Cooper, D.L.; Lovett, S.T. Recombinational branch migration by the RadA/Sms paralog of RecA in Escherichia coli. *Elife* **2016**, *5*. [CrossRef] [PubMed]

112. Marie, L.; Rapisarda, C.; Morales, V.; Berge, M.; Perry, T.; Soulet, A.L.; Gruget, C.; Remaut, H.; Fronzes, R.; Polard, P. Bacterial RadA is a DnaB-type helicase interacting with RecA to promote bidirectional D-loop extension. *Nat. Commun.* **2017**, *8*, 15638. [CrossRef] [PubMed]

113. Sperschneider, J.; Catanzariti, A.-M.; DeBoer, K.; Petre, B.; Gardiner, D.M.; Singh, K.B.; Dodds, P.N.; Taylor, J.M. LOCALIZER: Subcellular localization prediction of both plant and effector proteins in the plant cell. *Sci. Rep.* **2017**, *7*, 44598. [CrossRef] [PubMed]

114. Sievers, F.; Wilm, A.; Dineen, D.; Gibson, T.J.; Karplus, K.; Li, W.; Lopez, R.; McWilliam, H.; Remmert, M.; Soding, J.; et al. Fast, scalable generation of high-quality protein multiple sequence alignments using Clustal Omega. *Mol. Syst. Biol.* **2011**, *7*, 539. [CrossRef]

115. Torres, R.; Serrano, E.; Alonso, J.C. Bacillus subtilis RecA interacts with and loads RadA/Sms to unwind recombination intermediates during natural chromosomal transformation. *Nucleic Acids Res.* **2019**, *47*, 9198–9215. [CrossRef]

116. Whitby, M.C.; Ryder, L.; Lloyd, R.G. Reverse branch migration of Holliday junctions by RecG protein: A new mechanism for resolution of intermediates in recombination and DNA repair. *Cell* **1993**, *75*, 341–350. [CrossRef]

117. McGlynn, P.; Lloyd, R.G. RecG helicase activity at three- and four-strand DNA structures. *Nucleic Acids Res.* **1999**, *27*, 3049–3056. [CrossRef]

118. Warren, G.M.; Stein, R.A.; McHaourab, H.S.; Eichman, B.F. Movement of the RecG Motor Domain upon DNA Binding Is Required for Efficient Fork Reversal. *Int. J. Mol. Sci.* **2018**, *19*, 3049. [CrossRef]

119. Azeroglu, B.; Mawer, J.S.; Cockram, C.A.; White, M.A.; Hasan, A.M.; Filatenkova, M.; Leach, D.R. RecG Directs DNA Synthesis during Double-Strand Break Repair. *PLoS Genet.* **2016**, *12*, e1005799. [CrossRef]

120. Bianco, P.R.; Lyubchenko, Y.L. SSB and the RecG DNA helicase: An intimate association to rescue a stalled replication fork. *Protein Sci.* **2017**, *26*, 638–649. [CrossRef]

121. Wallet, C.; Le Ret, M.; Bergdoll, M.; Bichara, M.; Dietrich, A.; Gualberto, J.M. The RECG1 DNA Translocase Is a Key Factor in Recombination Surveillance, Repair, and Segregation of the Mitochondrial DNA in Arabidopsis. *Plant Cell* **2015**, *27*, 2907–2925. [CrossRef] [PubMed]

122. Takeuchi, R.; Kimura, S.; Saotome, A.; Sakaguchi, K. Biochemical properties of a plastidial DNA polymerase of rice. *Plant Mol. Biol.* **2007**, *64*, 601–611. [CrossRef] [PubMed]

123. Moriyama, T.; Terasawa, K.; Sato, N. Conservation of POPs, the plant organellar DNA polymerases, in eukaryotes. *Protist* **2011**, *162*, 177–187. [CrossRef] [PubMed]

124. Moriyama, T.; Sato, N. Enzymes involved in organellar DNA replication in photosynthetic eukaryotes. *Front. Plant Sci.* **2014**, *5*, 480. [CrossRef]

125. Mori, Y.; Kimura, S.; Saotome, A.; Kasai, N.; Sakaguchi, N.; Uchiyama, Y.; Ishibashi, T.; Yamamoto, T.; Chiku, H.; Sakaguchi, K. Plastid DNA polymerases from higher plants, Arabidopsis thaliana. *Biochem. Biophys. Res. Commun.* **2005**, *334*, 43–50. [CrossRef]

126. Zaegel, V.; Guermann, B.; Le Ret, M.; Andres, C.; Meyer, D.; Erhardt, M.; Canaday, J.; Gualberto, J.M.; Imbault, P. The plant-specific ssDNA binding protein OSB1 is involved in the stoichiometric transmission of mitochondrial DNA in Arabidopsis. *Plant Cell* **2006**, *18*, 3548–3563. [CrossRef]

127. Abdelnoor, R.V.; Christensen, A.C.; Mohammed, S.; Munoz-Castillo, B.; Moriyama, H.; Mackenzie, S.A. Mitochondrial genome dynamics in plants and animals: Convergent gene fusions of a MutS homologue. *J. Mol. Evol.* **2006**, *63*, 165–173. [CrossRef]

128. Desveaux, D.; Allard, J.; Brisson, N.; Sygusch, J. A new family of plant transcription factors displays a novel ssDNA-binding surface. *Nat. Struct. Biol.* **2002**, *9*, 512–517. [CrossRef]

129. Foyer, C.H.; Karpinska, B.; Krupinska, K. The functions of WHIRLY1 and REDOX-RESPONSIVE TRANSCRIPTION FACTOR 1 in cross tolerance responses in plants: A hypothesis. *Philos. Trans. R. Soc. B Biol. Sci.* **2014**, *369*, 20130226. [CrossRef]

130. Beltran, J.; Wamboldt, Y.; Sanchez, R.; LaBrant, E.W.; Kundariya, H.; Virdi, K.S.; Elowsky, C.; Mackenzie, S.A. Specialized Plastids Trigger Tissue-Specific Signaling for Systemic Stress Response in Plants. *Plant Physiol.* **2018**, *178*, 672–683. [CrossRef]

131. Chan, Y.W.; Mohr, R.; Millard, A.D.; Holmes, A.B.; Larkum, A.W.; Whitworth, A.L.; Mann, N.H.; Scanlan, D.J.; Hess, W.R.; Clokie, M.R. Discovery of cyanophage genomes which contain mitochondrial DNA polymerase. *Mol. Biol. Evol.* **2011**, *28*, 2269–2274. [CrossRef] [PubMed]

132. Ayala-Garcia, V.M.; Baruch-Torres, N.; Garcia-Medel, P.L.; Brieba, L.G. Plant organellar DNA polymerases paralogs exhibit dissimilar nucleotide incorporation fidelity. *FEBS J.* **2018**, *285*, 4005–4018. [CrossRef] [PubMed]

133. Trasvina-Arenas, C.H.; Baruch-Torres, N.; Cordoba-Andrade, F.J.; Ayala-Garcia, V.M.; Garcia-Medel, P.L.; Diaz-Quezada, C.; Peralta-Castro, A.; Ordaz-Ortiz, J.J.; Brieba, L.G. Identification of a unique insertion in plant organellar DNA polymerases responsible for 5′-dRP lyase and strand-displacement activities: Implications for Base Excision Repair. *DNA Repair* **2018**, *65*, 1–10. [CrossRef] [PubMed]

134. Baruch-Torres, N.; Brieba, L.G. Plant organellar DNA polymerases are replicative and translesion DNA synthesis polymerases. *Nucleic Acids Res.* **2017**, *45*, 10751–10763. [CrossRef] [PubMed]

135. Parent, J.S.; Lepage, E.; Brisson, N. Divergent roles for the two PolI-like organelle DNA polymerases of Arabidopsis. *Plant Physiol.* **2011**, *156*, 254–262. [CrossRef] [PubMed]

136. Morley, S.A.; Nielsen, B.L. Chloroplast DNA Copy Number Changes during Plant Development in Organelle DNA Polymerase Mutants. *Front. Plant Sci.* **2016**, *7*, 57. [CrossRef]

137. Boshoff, H.I.; Reed, M.B.; Barry, C.E., 3rd; Mizrahi, V. DnaE2 polymerase contributes to in vivo survival and the emergence of drug resistance in Mycobacterium tuberculosis. *Cell* **2003**, *113*, 183–193. [CrossRef]

138. Obeid, S.; Schnur, A.; Gloeckner, C.; Blatter, N.; Welte, W.; Diederichs, K.; Marx, A. Learning from directed evolution: Thermus aquaticus DNA polymerase mutants with translesion synthesis activity. *Chembiochem* **2011**, *12*, 1574–1580. [CrossRef]

139. Despres, C.; Subramaniam, R.; Matton, D.P.; Brisson, N. The Activation of the Potato PR-10a Gene Requires the Phosphorylation of the Nuclear Factor PBF-1. *Plant Cell* **1995**, *7*, 589–598. [CrossRef]

140. Cappadocia, L.; Parent, J.S.; Zampini, E.; Lepage, E.; Sygusch, J.; Brisson, N. A conserved lysine residue of plant Whirly proteins is necessary for higher order protein assembly and protection against DNA damage. *Nucleic Acids Res.* **2012**, *40*, 258–269. [CrossRef]
141. Cappadocia, L.; Marechal, A.; Parent, J.S.; Lepage, E.; Sygusch, J.; Brisson, N. Crystal structures of DNA-Whirly complexes and their role in Arabidopsis organelle genome repair. *Plant Cell* **2010**, *22*, 1849–1867. [CrossRef] [PubMed]
142. Krause, K.; Kilbienski, I.; Mulisch, M.; Rodiger, A.; Schafer, A.; Krupinska, K. DNA-binding proteins of the Whirly family in Arabidopsis thaliana are targeted to the organelles. *FEBS Lett.* **2005**, *579*, 3707–3712. [CrossRef] [PubMed]
143. Marechal, A.; Parent, J.S.; Veronneau-Lafortune, F.; Joyeux, A.; Lang, B.F.; Brisson, N. Whirly proteins maintain plastid genome stability in Arabidopsis. *Proc. Natl. Acad. Sci. USA* **2009**, *106*, 14693–14698. [CrossRef] [PubMed]
144. Marechal, A.; Parent, J.S.; Sabar, M.; Veronneau-Lafortune, F.; Abou-Rached, C.; Brisson, N. Overexpression of mtDNA-associated AtWhy2 compromises mitochondrial function. *BMC Plant Biol.* **2008**, *8*, 42. [CrossRef] [PubMed]
145. Rédei, G.P.; Plurad, S.B. Hereditary Structural Alterations of Plastids Induced by a Nuclear Mutator Gene in Arabidopsis. *Protoplasma* **1973**, *77*, 361–380.
146. Rédei, G.P. Extra-chromosomal mutability determined by a nuclear gene locus in Arabidopsis. *Mutat. Res.* **1973**, *18*, 149–162. [CrossRef]
147. Martinez-Zapater, J.M.; Gil, P.; Capel, J.; Somerville, C.R. Mutations at the Arabidopsis CHM locus promote rearrangements of the mitochondrial genome. *Plant Cell* **1992**, *4*, 889–899. [CrossRef]
148. Modrich, P.; Lahue, R. Mismatch repair in replication fidelity, genetic recombination, and cancer biology. *Annu. Rev. Biochem.* **1996**, *65*, 101–133. [CrossRef]
149. Brockman, S.A.; McFadden, C.S. The mitochondrial genome of Paraminabea aldersladei (Cnidaria: Anthozoa: Octocorallia) supports intramolecular recombination as the primary mechanism of gene rearrangement in octocoral mitochondrial genomes. *Genome Biol. Evol.* **2012**, *4*, 994–1006. [CrossRef]
150. Figueroa, D.F.; Baco, A.R. Octocoral mitochondrial genomes provide insights into the phylogenetic history of gene order rearrangements, order reversals, and cnidarian phylogenetics. *Genome Biol. Evol.* **2014**, *7*, 391–409. [CrossRef]
151. Fukui, K.; Harada, A.; Wakamatsu, T.; Minobe, A.; Ohshita, K.; Ashiuchi, M.; Yano, T. The GIY-YIG endonuclease domain of Arabidopsis MutS homolog 1 specifically binds to branched DNA structures. *FEBS Lett.* **2018**, *592*, 4066–4077. [CrossRef] [PubMed]
152. Pinto, A.V.; Mathieu, A.; Marsin, S.; Veaute, X.; Ielpi, L.; Labigne, A.; Radicella, J.P. Suppression of homologous and homeologous recombination by the bacterial MutS2 protein. *Mol. Cell* **2005**, *17*, 113–120. [CrossRef] [PubMed]
153. Damke, P.P.; Dhanaraju, R.; Marsin, S.; Radicella, J.P.; Rao, D.N. The nuclease activities of both the Smr domain and an additional LDLK motif are required for an efficient anti-recombination function of Helicobacter pylori MutS2. *Mol. Microbiol.* **2015**, *96*, 1240–1256. [CrossRef] [PubMed]
154. Wall, M.K.; Mitchenall, L.A.; Maxwell, A. Arabidopsis thaliana DNA gyrase is targeted to chloroplasts and mitochondria. *Proc. Natl. Acad. Sci. USA* **2004**, *101*, 7821–7826. [CrossRef]
155. Evans-Roberts, K.M.; Mitchenall, L.A.; Wall, M.K.; Leroux, J.; Mylne, J.S.; Maxwell, A. DNA Gyrase Is the Target for the Quinolone Drug Ciprofloxacin in Arabidopsis thaliana. *J. Biol. Chem.* **2016**, *291*, 3136–3144. [CrossRef]
156. Soczek, K.M.; Grant, T.; Rosenthal, P.B.; Mondragon, A. CryoEM structures of open dimers of gyrase A in complex with DNA illuminate mechanism of strand passage. *Elife* **2018**, *7*. [CrossRef]
157. Papillon, J.; Menetret, J.F.; Batisse, C.; Helye, R.; Schultz, P.; Potier, N.; Lamour, V. Structural insight into negative DNA supercoiling by DNA gyrase, a bacterial type 2A DNA topoisomerase. *Nucleic Acids Res.* **2013**, *41*, 7815–7827. [CrossRef]
158. Chen, S.F.; Huang, N.L.; Lin, J.H.; Wu, C.C.; Wang, Y.R.; Yu, Y.J.; Gilson, M.K.; Chan, N.L. Structural insights into the gating of DNA passage by the topoisomerase II DNA-gate. *Nat. Commun.* **2018**, *9*, 3085. [CrossRef]
159. Drlica, K.; Malik, M.; Kerns, R.J.; Zhao, X. Quinolone-mediated bacterial death. *Antimicrob. Agents Chemother.* **2008**, *52*, 385–392. [CrossRef]

160. Wentzell, L.M.; Maxwell, A. The complex of DNA gyrase and quinolone drugs on DNA forms a barrier to the T7 DNA polymerase replication complex. *J. Mol. Biol.* **2000**, *304*, 779–791. [CrossRef]

161. Stracy, M.; Wollman, A.J.M.; Kaja, E.; Gapinski, J.; Lee, J.E.; Leek, V.A.; McKie, S.J.; Mitchenall, L.A.; Maxwell, A.; Sherratt, D.J.; et al. Single-molecule imaging of DNA gyrase activity in living Escherichia coli. *Nucleic Acids Res.* **2019**, *47*, 210–220. [CrossRef] [PubMed]

162. Waterworth, W.M.; Kozak, J.; Provost, C.M.; Bray, C.M.; Angelis, K.J.; West, C.E. DNA ligase 1 deficient plants display severe growth defects and delayed repair of both DNA single and double strand breaks. *BMC Plant Biol.* **2009**, *9*, 79. [CrossRef] [PubMed]

163. Sunderland, P.A.; West, C.E.; Waterworth, W.M.; Bray, C.M. An evolutionarily conserved translation initiation mechanism regulates nuclear or mitochondrial targeting of DNA ligase 1 in Arabidopsis thaliana. *Plant J.* **2006**, *47*, 356–367. [CrossRef] [PubMed]

164. Ellenberger, T.; Tomkinson, A.E. Eukaryotic DNA ligases: Structural and functional insights. *Annu. Rev. Biochem.* **2008**, *77*, 313–338. [CrossRef] [PubMed]

165. Willer, M.; Rainey, M.; Pullen, T.; Stirling, C.J. The yeast CDC9 gene encodes both a nuclear and a mitochondrial form of DNA ligase I. *Curr. Biol.* **1999**, *9*, 1085–1094. [CrossRef]

166. Pascal, J.M.; O'Brien, P.J.; Tomkinson, A.E.; Ellenberger, T. Human DNA ligase I completely encircles and partially unwinds nicked DNA. *Nature* **2004**, *432*, 473–478. [CrossRef]

167. Howes, T.R.; Tomkinson, A.E. DNA ligase I, the replicative DNA ligase. *Subcell. Biochem.* **2012**, *62*, 327–341. [CrossRef]

168. Simsek, D.; Furda, A.; Gao, Y.; Artus, J.; Brunet, E.; Hadjantonakis, A.K.; Van Houten, B.; Shuman, S.; McKinnon, P.J.; Jasin, M. Crucial role for DNA ligase III in mitochondria but not in Xrcc1-dependent repair. *Nature* **2011**, *471*, 245–248. [CrossRef]

169. Bianchi, J.; Rudd, S.G.; Jozwiakowski, S.K.; Bailey, L.J.; Soura, V.; Taylor, E.; Stevanovic, I.; Green, A.J.; Stracker, T.H.; Lindsay, H.D.; et al. PrimPol bypasses UV photoproducts during eukaryotic chromosomal DNA replication. *Mol. Cell* **2013**, *52*, 566–573. [CrossRef]

170. Garcia-Gomez, S.; Reyes, A.; Martinez-Jimenez, M.I.; Chocron, E.S.; Mouron, S.; Terrados, G.; Powell, C.; Salido, E.; Mendez, J.; Holt, I.J.; et al. PrimPol, an archaic primase/polymerase operating in human cells. *Mol. Cell* **2013**, *52*, 541–553. [CrossRef]

171. Wan, L.; Lou, J.; Xia, Y.; Su, B.; Liu, T.; Cui, J.; Sun, Y.; Lou, H.; Huang, J. hPrimpol1/CCDC111 is a human DNA primase-polymerase required for the maintenance of genome integrity. *EMBO Rep.* **2013**, *14*, 1104–1112. [CrossRef] [PubMed]

172. Rechkoblit, O.; Gupta, Y.K.; Malik, R.; Rajashankar, K.R.; Johnson, R.E.; Prakash, L.; Prakash, S.; Aggarwal, A.K. Structure and mechanism of human PrimPol, a DNA polymerase with primase activity. *Sci. Adv.* **2016**, *2*, e1601317. [CrossRef] [PubMed]

173. Keen, B.A.; Jozwiakowski, S.K.; Bailey, L.J.; Bianchi, J.; Doherty, A.J. Molecular dissection of the domain architecture and catalytic activities of human PrimPol. *Nucleic Acids Res.* **2014**, *42*, 5830–5845. [CrossRef] [PubMed]

174. Martinez-Jimenez, M.I.; Calvo, P.A.; Garcia-Gomez, S.; Guerra-Gonzalez, S.; Blanco, L. The Zn-finger domain of human PrimPol is required to stabilize the initiating nucleotide during DNA priming. *Nucleic Acids Res.* **2018**, *46*, 4138–4151. [CrossRef] [PubMed]

175. Bailey, L.J.; Bianchi, J.; Hegarat, N.; Hochegger, H.; Doherty, A.J. PrimPol-deficient cells exhibit a pronounced G2 checkpoint response following UV damage. *Cell Cycle* **2016**, *15*, 908–918. [CrossRef] [PubMed]

176. Kopinski, P.K.; Janssen, K.A.; Schaefer, P.M.; Trefely, S.; Perry, C.E.; Potluri, P.; Tintos-Hernandez, J.A.; Singh, L.N.; Karch, K.R.; Campbell, S.L.; et al. Regulation of nuclear epigenome by mitochondrial DNA heteroplasmy. *Proc. Natl. Acad. Sci. USA* **2019**, *116*, 16028–16035. [CrossRef]

 *plants*

Review

Factors Affecting Organelle Genome Stability in *Physcomitrella patens*

Masaki Odahara

Biomacromolecules Research Team, RIKEN Center for Sustainable Resource Science, 2-1 Hirosawa, Wako-shi, Saitama 351-0198, Japan; masaki.odahara@riken.jp; Tel.: +81-48-462-1111

Received: 16 December 2019; Accepted: 21 January 2020; Published: 23 January 2020

Abstract: Organelle genomes are essential for plants; however, the mechanisms underlying the maintenance of organelle genomes are incompletely understood. Using the basal land plant *Physcomitrella patens* as a model, nuclear-encoded homologs of bacterial-type homologous recombination repair (HRR) factors have been shown to play an important role in the maintenance of organelle genome stability by suppressing recombination between short dispersed repeats. In this review, I summarize the factors and pathways involved in the maintenance of genome stability, as well as the repeats that cause genomic instability in organelles in *P. patens*, and compare them with findings in other plant species. I also discuss the relationship between HRR factors and organelle genome structure from the evolutionary standpoint.

Keywords: chloroplast; mitochondrion; genome stability; homologous recombination repair; repeated sequence; *Physcomitrella patens*

---

1. Introduction

Physcomitrella patens is a moss (bryophyte) that has been used as a model species for studying cell growth and differentiation [1]. Additionally, *P. patens* is recognized as a model for land plants because it is located at the base of the land plant lineage [2]. The life cycle of *P. patens* is simple and mostly haploid. Germinated spores of *P. patens* produce filamentous protonemal cells comprising chloronemal and caulonemal cells, which subsequently produce gametophores with leafy shoots. Sporophyte, the only diploid phase in the life cycle of *P. patens*, is developed from zygotes, archegonia, and antheridia, which are formed at the top of gametophores. Nuclear DNA of *P. patens* shows exceptionally high activity of homologous recombination, which enables its use for gene targeting in combination with polyethylene glycol-mediated protoplast transformation [3]. This feature, together with its haploid vegetative growth phase and recent advances in nuclear genome analysis, has accelerated reverse genetic analyses in *P. patens* [2,4].

Each *P. patens* cell harbors ≈50 large spindle-shaped chloroplasts and many rod- or sphere-shaped mitochondria. Chloroplast and mitochondria in *P. patens*, as in other plant species and algae, possess their own DNA, which associates with proteins to form nucleoids. The mitochondrial DNA (mtDNA) of *P. patens* is 105 kb in size and harbors genes encoding transfer RNAs (tRNAs), ribosomal RNAs (rRNAs), and proteins that regulate gene expression and oxidative phosphorylation [5]. The mapped mitochondrial genomes of angiosperms are larger than that of *P. patens*; however, they are shown to form complicated structures including linear, branched, and circular structures [6]. Moreover, homologous recombination between repeats longer than 1 kb, which are frequently observed in angiosperm mtDNA, makes them a more complicated structure. By contrast, *P. patens* mtDNA forms a single circular structure because of the absence of repeats longer than 80 bp [5,7–9]. The chloroplast DNA (cpDNA) of *P. patens* is 123 kb in size and contains genes encoding tRNAs, rRNAs, and proteins including subunits of RNA polymerase- and photosynthesis-related proteins [10]. The cpDNA of *P. patens* exhibits a typical circular structure with large single-copy (LSC) and small single-copy (SSC) regions separated by a pair of large inverted repeat (IR) regions [10]. Except for the large IR regions (9.6 kb each), the longest dispersed repeat in *P. patens* cpDNA is 63 bp in size, with a 3 bp mismatch [11]. Notably, neither mtDNA nor cpDNA encode proteins that are involved in DNA replication, recombination, and repair; instead, proteins involved in these processes are encoded by nuclear DNA, similar to a large number of proteins that function in chloroplasts and mitochondria.

2. Plant Homologs of Bacterial Proteins and Their Localization

Because chloroplasts and mitochondria are derived from bacteria, internal contents of these organelles resemble prokaryotes. Although orthologs of bacterial proteins function in chloroplasts and mitochondria, most of the chloroplast and mitochondrial proteins are encoded by nuclear DNA because of gene transfer during evolution. In bacteria, homologous recombination repair (HRR) proteins repair DNA double-strand breaks and collapsed or stalled replication forks. Homologs of bacterial HRR factors are also found in the nuclear genome of *P. patens* and that of other plant species. The N-terminus of HRR factors contain signal peptides that target these proteins to chloroplasts and/or mitochondria. Interestingly, such bacterial-type HRR factors have not been found in animal or yeast nuclear genomes [8,12–14], implying the existence of plant-specific mechanisms underlying organelle DNA maintenance by HRR. Table 1 summarizes plant homologs of bacterial HRR factors and MutS homolog 1 (MSH1; involved in organelle genome stabilization) in *P. patens* and other plant species, including *Chlamydomonas reinhardtii* and *Arabidopsis thaliana*, which are representative models of green algae and angiosperms, respectively. Nuclear genomes of *P. patens* and other plant species encode several homologs of bacterial HRR factors, although some homologs have not been identified in the genomes of *P. patens* and other plant species, on the basis of sequence similarity.

Table 1. Summary of homologous recombination repair (HRR) factors and MutS homolog 1 (MSH1) in *Escherichia coli* and their plant homologs.

E. coli		*Physcomitrella patens*	*Arabidopsis thaliana*	*Chlamydomonas reinhardtii*
Protein	**Function [15–17]**	**Protein/Localization**		
RecFOR	Single-stranded DNA (ssDNA) binding RecA loading	-	-	-
RecBCD	DNA Helicase/exonuclease RecA loading	-	-	-
RecA	Homology search Strand exchange	RECA1/mt [13] RECA2/cp [18]	RECA1/cp [19] RECA2/cp, mt [21] RECA3/mt [22]	REC1/cp [20]
RecX	RecA regulation	RECX/cp, mt [8]	RECX	RECX
RecG	DNA Helicase/translocase	RECG/cp, mt [14]	RECG1/cp, mt [23]	-
RuvAB	Branch migration	-	-	-
RuvC	Holiday junction resolution	MOC1	MOC1/cp [24]	MOC1/cp [24]
MutS	Mismatch recognition	MSH1A/cp, mt [25] MSH1B/cp, mt [25]	MSH1/cp, mt [26]	MSH1

RecA is a key factor in HRR, as it binds to single-stranded DNA (ssDNA) and identifies homologous sequences to perform strand exchange between them [27]. Nuclear DNA of *P. patens* encodes two types of RecA homologs, RECA1 and RECA2, which show moderate sequence similarity. Phylogenetic analysis shows that these two RECA proteins cluster with either cyanobacterial RecA or proteobacterial RecA in separate clades, suggesting that these proteins have different origins, that is, RECA1 from α-proteobacteria, and RECA2 from cyanobacteria [13]. Products of *RECA1* and *RECA2* genes expressed from the nuclear DNA are predominantly localized to mitochondria and chloroplasts, respectively, thus reflecting their predicted origins [13,18]. When full-length RECA1 and RECA2 proteins are transiently produced in protoplasts, they form granular structures that associate with organelle nucleoids [8,18], indicating that these proteins constantly associate with and/or act on nucleoids. Consistent with this hypothesis, chloroplast RecA is shown to associate with the chloroplast nucleoid by nucleoids enriched proteome in maize [28]. Interestingly, although HRR factors are encoded by a single conserved gene in plants, the copy number of *RECA* varies among plant species. Although *A. thaliana* and other flowering plants harbor multiple copies of the *RECA* gene, and the encoded proteins localize to chloroplasts and/or mitochondria, algae, including *C. reinhardtii*, harbor a single *RECA* gene copy, and the encoded RecA homolog localizes to chloroplasts [12] (Table 1).

RecG, a DNA helicase/translocase, functions in the rescue of branched DNA structures including stalled replication forks [29]. The nuclear genome of *P. patens* harbors a single copy of the *RECG* gene [14]. Phylogenetic analysis shows that plant RecG homologs, including *P. patens* RECG, are closely related to cyanobacterial RecG, suggesting that these proteins originated from cyanobacteria [23]. The RECG protein of *P. patens* harbors an ambiguous N-terminal signal peptide but localizes to both chloroplasts and mitochondria, similar to the *A. thaliana* RecG homolog, RECG1 [14,23]. Moreover, full-length *P. patens* RECG protein localizes to nucleoids of both organelles [14].

Unlike RecA and RecG, RecX does not act directly on DNA but participates in HRR by directly regulating RecA activity [30]. Although RecX is absent from several bacterial classes including α-proteobacteria and cyanobacteria [31], it is encoded by single copy genes present in the nuclear genomes of diverse plants ranging from green algae to angiosperms [8]. Because of difficulty in analyzing the evolutionary origin of plant RecX homologs, it is unclear whether α-proteobacteria and cyanobacteria lost their RecX or plants acquired RecX via horizontal gene transfer. In protoplasts, a fluorescent protein-tagged RecX homolog of *P. patens*, RECX, localizes to mitochondrial and chloroplast nucleoids, thereby co-localizing with RECA1 and RECA2, respectively [8].

MSH is a eukaryotic homolog of bacterial MutS. Among several types of MSH proteins, MSH1 is the only protein that localizes to organelles [32,33]. MSH1 was originally identified in *A. thaliana* as a chloroplast mutator (CHM) protein because of the variegated phenotype of the mutant [34,35]. MSH1 is distinct from other MSH proteins and MutS because of the presence of the GIY-YIG endonuclease domain at its C-terminal end [21]. The nuclear genome of *P. patens* harbors two *MSH1* genes, *MSH1A* and *MSH1B*, although nuclear genomes of other plants carry only one *MSH1* gene copy. Because MSH1A lacks the C-terminal endonuclease domain, *P. patens MSH1* genes are thought to be derived by gene duplication or the loss of C-termini endonuclease domains after the duplication event [25]. Both *P. patens* MSH1 proteins (MSH1A and MSH1B) localize to organelle nucleoids by forming granular structures [25], similar to the MSH1 localization pattern in *A. thaliana* [26].

3. Maintenance of Mitochondrial Genome Stability by HRR and MSH1

3.1. RECA

P. patens mitochondrial *RECA1* knockout (KO) mutants generated by targeted gene disruption show severe defects in protonema cells, with less-developed gametophores and defective mitochondria characterized by an enlarged shape, disorganized cristae, and lower matrix electron density [7], indicating that *RECA1* is essential for normal growth. The mitochondrial genome of *P. patens RECA1* KO mutant is destabilized by the accumulation of products derived from aberrant recombination between short repeats dispersed throughout the mtDNA [7]. Most of the 24 pairs of repeats (≥30 bp) identified in *P. patens* mtDNA are involved in recombination in *RECA1* KO plants [8], occasionally leading to the generation of subgenomes [7]. Interestingly, because most of the repeats are located in introns of genes in the direct orientation, recombination between them leads to the loss of genes and generation of subgenomes, which may be subsequently lost, as these are not replicated. Thus, copy number variation of loci resulting from the loss of subgenomes is associated with instability of mtDNA in the *RECA1* KO mutant [14]. Collectively, these findings show the role of RECA1 in maintaining mtDNA stability by suppressing aberrant recombination between short dispersed repeats (SDRs) in *P. patens*. Additionally, defects in the recovery of mtDNA damaged by methyl methanesulfonate (MMS) in *RECA1* KO plants suggest the involvement of RECA1 in the repair of exogenously damaged mtDNA [13].

In *A. thaliana*, two RecA homologs, RECA2 and RECA3, localize to mitochondria (Table 1). In comparison with RECA2, RECA3 is more diverged from other RECAs and has truncated C-terminus, which is considered unusual because the C-terminus of RecA is important for its function [21,36]. Consistent with the gene structure, *A. thaliana RECA2* mutants are seedling-lethal, thus indicating the importance of RECA2 for normal plant growth; by contrast, *RECA3* mutants are almost indistinguishable from the wild type [21]. Both *RECA2* and *RECA3* mutants accumulate products derived from recombination between intermediate-sized (100–300 bp) repeats in mtDNA, and the number of repeats involving recombination in *RECA2* mutants exceed that of *RECA3* mutants [36]. Although recombination between shorter repeats (<100 bp) has not been tested in *A. thaliana RECA2* and *RECA3* mutants, the aforementioned findings suggest a fundamental role of plant mitochondrial RecA homologs in maintaining mitochondrial genome stability by suppressing aberrant recombination between short repeats.

3.2. RECG

KO mutation of *P. patens RECG* gene leads to growth and morphological defects that are similar to but milder than those caused by the KO mutation of *RECA1* in plants [14]. The *RECG* KO mutant plants exhibit abnormal mitochondria, with disorganized cristae and lower matrix density. Moreover, mtDNA of the *RECG* KO mutant is destabilized by SDR-mediated recombination, similar to the mtDNA of the *RECA1* KO mutant, and the length of repeats involved in recombination is also similar between *RECA1* and *RECG* KO mutants [14]. However, these repeats exhibit some differences between *RECA1* and *RECG* KO mutants; for example, at the mitochondrial *atp9* locus, recombination between *ccmF* and *atp9* mediated by 47 bp repeats leads to product accumulation in mitochondria of the *RECG* KO mutant, whereas recombination between *nad2* and *atp9* mediated by 60 bp repeats, which is a hallmark of recombination induced by the *RECA1* KO mutation [7], does not lead to product accumulation in mitochondria of the *RECG* KO mutant [14]. Furthermore, increase in copy numbers of all tested loci in the *RECG* KO mutant differed from that in the *RECA1* KO mutant. These differences suggest that RECG of *P. patens* plays a somewhat different role from that of RECA1 in the maintenance of mtDNA stability. Because the amount of mitochondrial recombination products often show a direct correlation with the heterogeneous *RECG* KO growth defects, recombination between mitochondrial SDRs is considered as the cause of all morphological phenotypes [14]. Because of mtDNA rearrangements induced by the KO mutation of *RECG*, the level of mitochondrial transcripts is decreased by recombination between repeats located in introns of mitochondrial genes [14]. Although *A. thaliana RECG1* mutants are morphologically indistinguishable from wild-type plants under normal growth conditions, they show mtDNA instability because of aberrant recombination between intermediate-sized repeats (100–500 bp in length) [23]. Thus, RECG1 participates in the suppression of recombination between intermediate-sized repeats, and the loss of *RECG1* leading to the accumulation of recombination products. Although recombination between shorter repeats has not been analyzed in *A. thaliana RECG1* mutants, recombination surveillance indicates that RecG homolog is involved in the suppression of aberrant recombination between short and/or imperfect repeats in plant mitochondria.

3.3. RECX

KO mutation of *P. patens RECX*, which leads to no significant morphological phenotypes, results in a minor but reliable increase in products derived from recombination between several pairs of mitochondrial SDRs [8], suggesting the involvement of RECX in the maintenance of mtDNA stability. Overexpression (OEX) of *P. patens RECX* in plants leads to mtDNA instability because of the induction of recombination between many pairs of SDRs, sometimes with a comparable level with mtDNA instability in the *RECA1* KO mutant [8]. Taking into account the protein–protein interaction between *P. patens* RECX and RECA1, as revealed by yeast two-hybrid assays, RECX is believed to modulate the function of RECA1 by directly binding to RECA1 to maintain mtDNA stability, rather than inducing mtDNA instability in wild type. The involvement of *RECX* in the maintenance of mtDNA stability is also supported by the positive correlation between the expression of *RECX* and other mtDNA stabilizing genes, including *RECA1* and *RECG*, in several tissues of *P. patens* [8]. Interestingly, the expression of *RECX*, *RECA1*, *RECG*, and *MSH1B* is highly increased in *P. patens* spores, thus indicating their roles in mtDNA maintenance during transmission to progenies.

3.4. MSH1

Because *P. patens* unusually possesses two *MSH1* genes, single and double KO mutants of *MSH1* genes were generated. Although the single and double *MSH1* mutants showed no significant phenotypes compared with the wild type, comparison among the mutants show an involvement of *MSH1B* in the maintenance of mtDNA [25]. In the single *MSH1B* KO mutant and *MSH1A* and *MSH1B* double KO mutants, mtDNA is similarly destabilized by the induction of recombination between mitochondrial repeats (21–69 bp in length) that overlap with those in *P. patens RECA1* or *RECG* KO mitochondria. On the other hand, the accumulation of products derived from recombination between *nad2* and *atp9*, rather than that of products derived from recombination between *ccmF* and *atp9*, hallmarks of the mitochondrial *atp9* locus in *RECA1* KO and *RECG* KO mutants, respectively, in the *MSH1B* mutant suggest a similar mechanism of mtDNA stabilization between MSH1B and RECA1, whereas the *MSH1 RECA1* double KO mutant is likely lethal [25]. Genetic interaction between *P. patens MSH1B* and *RECG* loci, as shown by epistatic analysis of the suppression of recombination, suggests that MSH1B and RECA1 act in distinct pathways that converge at a node in mitochondria [25]. The importance of the GIY-YIG endonuclease domain of MSH1 for the suppression of recombination is indicated by its deletion mutants; on the other hand, no significant phenotypes are observed in the *MSH1A* KO mutant, which lacks the endonuclease domain [25]. The instability of mtDNA in *A. thaliana MSH1* mutants is well characterized; in these mutants, recombination is observed between 50–556 bp repeats, and the length of these repeats overlaps with that of repeats responsible for mtDNA instability in the *P. patens MSH1B* KO mutant [21,32,37]. Moreover, the difference in mtDNA rearrangements between *A. thaliana MSH1* mutants and *RECA3* mutants, as well as the highly pronounced phenotypes of the *MSH1 RECA3* double KO mutants, suggest that these genes act in distinct but overlapping pathways [21]. Recent biochemical characterization of the GIY-YIG domain of *A. thaliana* MSH1 shows its binding to a branched DNA structure, proposing a mechanism for the suppression of recombination between repeats [38].

4. Maintenance of Chloroplast Genome Stability by HRR Proteins and MSH1

4.1. RECA

KO mutation of *P. patens RECA2* results in modest growth inhibition under glucose-deficient conditions and increased sensitivity to MMS or ultraviolet (UV) radiation, leading to DNA damage [11]. These phenotypes of the *RECA2* KO mutant are in contrast to those of the *RECA1* KO mutant of *P. patens*, which show severe growth defects under normal conditions. However, despite the slight effect of *RECA2* KO mutation on the morphology of *P. patens*, the cpDNA of the *RECA2* KO mutant is destabilized by the induction of recombination between SDRs (13–63 bp in length) [11]. This shows that

RECA2 is involved in the maintenance of chloroplast genome stability by suppressing recombination between SDRs. Moreover, roles of RECA1 and RECA2 in mitochondria and chloroplasts suggest the common role of RecA homologs in maintaining organelle genome stability by suppressing aberrant recombination between SDRs. Because *P. patens* cpDNA has fewer relatively long (>35 bp) repeats, the lack of RecA homologs may lead to a slight effect on the stability of cpDNA compared with that of mtDNA. Impaired recovery of damaged cpDNA, but not that of nuclear DNA or mtDNA, in *P. patens RECA2* KO mutants suggests another role of RECA2 in the maintenance of cpDNA stability by promoting recovery from DNA damage [11]. In contrast to the modest phenotypes of *P. patens* lacking chloroplast RECA, the deficiency of chloroplast RECA (RECA1) in *A. thaliana* plants (Table 1) is lethal [21]. *A. thaliana* T-DNA insertion *RECA1* mutants in which the level of *RECA1* transcripts is decreased to 15% of that in the wild type suggest that RECA1 is involved in the maintenance of cpDNA integrity by maintaining the quantity and multimeric structure of cpDNA [39]. *A. thaliana* RECA1 also maintains cpDNA stability by preventing cpDNA rearrangements in plants carrying a mutation in *Whirly* genes, which encode a family of ssDNA-binding proteins that suppress cpDNA rearrangements [40,41]. Chloroplast RECA in *C. reinhardtii* (Table 1) is also involved in the maintenance of chloroplast genome stability by suppressing aberrant recombination between SDRs, and it regulates the dynamics of chloroplast nucleoid including segregation [42].

4.2. RECG

Because the morphological defects of *RECG* KO mutant plants are similar to those of *RECA1* KO mutant plants, the defects of *RECG* KO plants are mainly attributed to defects in mtDNA. However, KO mutation of *RECG* leads to abnormal chloroplasts that over-accumulate starch and possess less-developed thylakoids, implying defects in chloroplast function [14]. Indeed, cpDNA and mtDNA of the *RECG* KO mutant are destabilized by the induction of recombination between SDRs. The repeats involved in recombination are almost common between the cpDNA of *RECG* and *RECA2* KO mutants, although the accumulation of recombination products is higher in the *RECG* KO mutant than in the *RECA2* KO mutant [14]. These results suggest that RECG maintains chloroplast genome stability by suppressing recombination between a broad range of repeats in cpDNA. Both synergistic and suppressive relationships are observed between *RECG* and *RECA2*, with respect to the suppression of recombination between chloroplast repeats, depending on the type of repeats [25], suggesting a complex relationship between these genes. Thus, *RECG* and *RECA2* may act in distinct pathways or in the same pathway, depending on the repeats, to suppress recombination. *A. thaliana* RECG1 localizes to chloroplasts; however, evidence indicating the involvement of RECG1 in the maintenance of chloroplast genome stability is lacking [23].

4.3. RECX

Although RECX localizes to chloroplast nucleoids, significant phenotypes have not been observed in the chloroplasts of *P. patens RECX* KO mutants and OEX plants. These KO and OEX plants show a basal level of products derived from recombination between chloroplast SDRs, in contrast to *P. patens RECA2* KO plants, which accumulate these recombinant products to high levels [8]. However, yeast two-hybrid assays show protein–protein interaction between *P. patens* RECX and RECA2, which is stronger than that between RECX and RECA1 [8]. This implies that RECX may interact with RECA2 and modulate its activity to maintain chloroplast genome stability, and the effect of *RECX* KO mutation or OEX was not evident probably because of the moderate effect of RECA2 inhibition on cpDNA.

4.4. MSH1

Similar to the instability of mitochondrial genome in the *MSH1* KO mutant, the *MSH1B* KO mutant shows chloroplast genome instability because of recombination between 28–63 bp SDRs in *P. patens* [25]. KO mutation of the *MSH1A* gene does not increase the abundance of recombination products in the wild-type or *MSH1B* KO mutant, indicating that *MSH1B* plays a predominant role in the suppression of recombination between SDRs in chloroplasts and mitochondria [25]. Interestingly, the level of recombination products in chloroplasts vary among the *P. patens MSH1B*, *RECA2*, and *RECG* KO mutant plants, depending on the type of repeats. Among these KO mutants, the level of products resulting from recombination between direct repeat-1 (DR-1) is the highest in *RECG* KO mutants, whereas the level of products resulting from recombination between inverted repeat-1 (IR-1) is the highest in *MSH1B* KO mutant plants [25]. This suggests a complicated regulation of recombination in chloroplasts. Similar complicated regulation is also observed in the genetic interaction between genes, as shown by synergistic relationships between *MSH1B* and *RECG* and between *MSH1B* and *RECA2*, although synergistic relationships have been observed for DR-1 but not for IR-1 [25]. Figure 1 summarizes all the factors affecting organelle stability and their relationship in *P. patens*. In *A. thaliana MSH1* mutants, cpDNA rearrangements at a locus containing a number of small repeats (<15 bp) indicate the involvement of MSH1 in maintaining chloroplast genome stability, although the details of these rearrangements remain unclear [26].

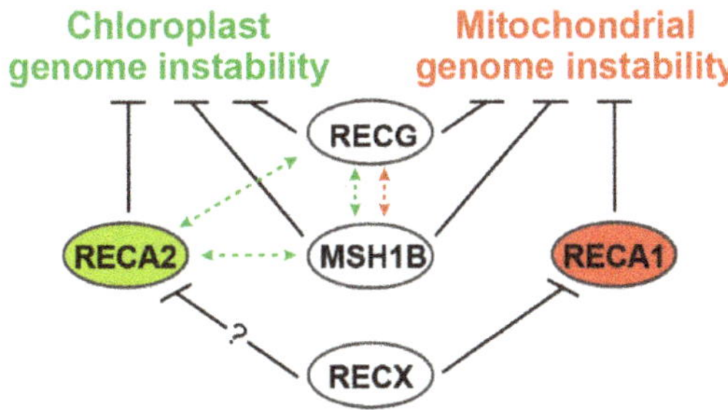

Figure 1. Factors affecting organelle genome stability in *P. patens*. Factors involving organelle genome stability are summarized with their relationship. Protein localization of the factors are shown by their colors: green (chloroplasts), red (mitochondria), and white (chloroplasts and mitochondria). Suppression and genetic relationship are shown by solid and dashed lines, respectively. RECX shows protein–protein interaction with RECA2, but its involvement in chloroplast genome stability remains unclear.

5. Organelle Genome Structure, Repeats, and HRR Proteins

Recent evidence in various plant species suggests the role of HRR factors in chloroplasts and mitochondria exclusively for the maintenance of genome stability by suppressing recombination between ectopic loci containing repeats, as summarized above. Because the phenomena of genome destabilization are common between mutants of organelle HRR factors, these factors likely function in a same suppression pathway. However, epistatic analyses of recombination suppression sometimes show that these factors act in distinct pathways [25]. Plant organelle HRR factors are thought to function in the repair of stalled or collapsed replication forks, which are prone to rearrangements in mutants [7]. Because such stalling and collapse of replication forks are caused by various types of DNA damage, the pathways of suppression in organelles may be regulated in a complicated manner. On the other hand, as shown in Table 1, not all HRR factors are conserved in plants, and some are absent in organelles of certain plant species; for example, mitochondrial RecA homologs are absent in some algae including *C. reinhardtii*, whereas copy numbers of mitochondrial RecA homologs are increased in various angiosperms including *A. thaliana* (Table 1) [12,13]. By contrast, chloroplast RecA copy numbers are conserved in plants (Table 1). Interestingly, the size and shape of mitochondrial genomes vary among plant species—*C. reinhardtii* possesses a 16 kb linear mitochondrial genome, whereas *A. thaliana* harbors a 368 kb multi-chromosome circular mitochondrial genome (Table 2). Moreover, the number of short repeats, which may lead to organelle genome instability because of the loss of HRR, corresponds to the size of the mitochondrial genome (Table 2). The presence/absence of RecA homologs may be correlated to the number and characteristics of repeats; RecA homologs are absent in algae because of the lack of significant repeats in mtDNA, whereas those in angiosperms are duplicated and functionally divergent to regulate recombination between increased and divergent repeats, or duplication of mitochondrial RecA homologs enabled increase of number of repeats in angiosperms. Recent advances in genome sequencing of various plant species provide an opportunity for exploring the relationship between HRR factors and organelle genome structure.

Table 2. Genome size and number of repeats in organelle.

Organelle	Feature	*C. reinhardtii*	*P. patens*	*A. thaliana*
Chloroplast	Genome size (bp)	203,828 [43]	122,890 [10]	154,478 [44]
	Number of repeats	>5000	55	31
Mitochondrion	Genome size (bp)	15,758 [45]	105,340 [5]	367,808 [46]
	Number of repeats	3	136	507

Repeats identified as ≥20 bp of direct or inverted repeats without mismatch by using REPuter [47].

Funding: This work was funded by SUMITOMO Foundation (170946) and the Japan Society for the Promotion of Science (19K22405).

Conflicts of Interest: The author declares no conflict of interest.

References

1. Vidali, L.; Bezanilla, M. Physcomitrella patens: A model for tip cell growth and differentiation. *Curr. Opin. Plant Biol.* **2012**, *15*, 625–631. [CrossRef] [PubMed]
2. Rensing, S.A.; Lang, D.; Zimmer, A.D.; Terry, A.; Salamov, A.; Shapiro, H.; Nishiyama, T.; Perroud, P.F.; Lindquist, E.A.; Kamisugi, Y.; et al. The Physcomitrella genome reveals evolutionary insights into the conquest of land by plants. *Science* **2008**, *319*, 64–69. [CrossRef] [PubMed]
3. Schaefer, D.G. A new moss genetics: Targeted mutagenesis in Physcomitrella patens. *Annu. Rev. Plant Biol.* **2002**, *53*, 477–501. [CrossRef] [PubMed]
4. Lang, D.; Ullrich, K.K.; Murat, F.; Fuchs, J.; Jenkins, J.; Haas, F.B.; Piednoel, M.; Gundlach, H.; Van Bel, M.; Meyberg, R.; et al. The Physcomitrella patens chromosome-scale assembly reveals moss genome structure and evolution. *Plant J.* **2018**, *93*, 515–533. [CrossRef] [PubMed]
5. Terasawa, K.; Odahara, M.; Kabeya, Y.; Kikugawa, T.; Sekine, Y.; Fujiwara, M.; Sato, N. The mitochondrial genome of the moss Physcomitrella patens sheds new light on mitochondrial evolution in land plants. *Mol. Biol. Evol.* **2007**, *24*, 699–709. [CrossRef]
6. Kozik, A.; Rowan, B.A.; Lavelle, D.; Berke, L.; Schranz, M.E.; Michelmore, R.W.; Christensen, A.C. The alternative reality of plant mitochondrial DNA: One ring does not rule them all. *PLoS Genet.* **2019**, *15*, e1008373. [CrossRef]
7. Odahara, M.; Kuroiwa, H.; Kuroiwa, T.; Sekine, Y. Suppression of Repeat-Mediated Gross Mitochondrial Genome Rearrangements by RecA in the Moss Physcomitrella patens. *Plant Cell* **2009**, *21*, 1182–1194. [CrossRef]
8. Odahara, M.; Sekine, Y. RECX Interacts with Mitochondrial RECA to Maintain Mitochondrial Genome Stability. *Plant Physiol.* **2018**, *177*, 300–310. [CrossRef]
9. Sloan, D.B. One ring to rule them all? Genome sequencing provides new insights into the 'master circle' model of plant mitochondrial DNA structure. *New Phytol.* **2013**, *200*, 978–985. [CrossRef]
10. Sugiura, C.; Kobayashi, Y.; Aoki, S.; Sugita, C.; Sugita, M. Complete chloroplast DNA sequence of the moss Physcomitrella patens: Evidence for the loss and relocation of rpoA from the chloroplast to the nucleus. *Nucleic Acids Res.* **2003**, *31*, 5324–5331. [CrossRef]
11. Odahara, M.; Inouye, T.; Nishimura, Y.; Sekine, Y. RECA plays a dual role in the maintenance of chloroplast genome stability in *Physcomitrella patens*. *Plant J.* **2015**, *84*, 516–526. [CrossRef] [PubMed]
12. Lin, Z.; Kong, H.; Nei, M.; Ma, H. Origins and evolution of the recA/RAD51 gene family: Evidence for ancient gene duplication and endosymbiotic gene transfer. *Proc. Natl. Acad. Sci. USA* **2006**, *103*, 10328–10333. [CrossRef] [PubMed]

13. Odahara, M.; Inouye, T.; Fujita, T.; Hasebe, M.; Sekine, Y. Involvement of mitochondrial-targeted RecA in the repair of mitochondrial DNA in the moss, *Physcomitrella patens*. *Genes Genet. Syst.* **2007**, *82*, 43–51. [CrossRef] [PubMed]

14. Odahara, M.; Masuda, Y.; Sato, M.; Wakazaki, M.; Harada, C.; Toyooka, K.; Sekine, Y. RECG Maintains Plastid and Mitochondrial Genome Stability by Suppressing Extensive Recombination between Short Dispersed Repeats. *PLoS Genet.* **2015**, *11*, e1005080. [CrossRef]

15. Kowalczykowski, S.C.; Dixon, D.A.; Eggleston, A.K.; Lauder, S.D.; Rehrauer, W.M. Biochemistry of homologous recombination in *Escherichia coli*. *Microbiol. Rev.* **1994**, *58*, 401–465. [CrossRef]

16. Venkatesh, R.; Ganesh, N.; Guhan, N.; Reddy, M.S.; Chandrasekhar, T.; Muniyappa, K. RecX protein abrogates ATP hydrolysis and strand exchange promoted by RecA: Insights into negative regulation of homologous recombination. *Proc. Natl. Acad. Sci. USA* **2002**, *99*, 12091–12096. [CrossRef]

17. Modrich, P.; Lahue, R. Mismatch repair in replication fidelity, genetic recombination, and cancer biology. *Annu. Rev. Biochem.* **1996**, *65*, 101–133. [CrossRef]

18. Inouye, T.; Odahara, M.; Fujita, T.; Hasebe, M.; Sekine, Y. Expression and Complementation Analyses of a Chloroplast-Localized Homolog of Bacterial RecA in the Moss *Physcomitrella patens*. *Biosci. Biotechnol. Biochem.* **2008**, *72*, 1340–1347. [CrossRef]

19. Cao, J.; Combs, C.; Jagendorf, A.T. The chloroplast-located homolog of bacterial DNA recombinase. *Plant Cell Physiol.* **1997**, *38*, 1319–1325. [CrossRef]

20. Nakazato, E.; Fukuzawa, H.; Tabata, S.; Takahashi, H.; Tanaka, K. Identification and Expression Analysis of cDNA Encoding a Chloroplast Recombination Protein REC1, the Chloroplast RecA Homologue in *Chlamydomonas reinhardtii*. *Biosci. Biotechnol. Biochem.* **2003**, *67*, 2608–2613. [CrossRef] [PubMed]

21. Shedge, V.; Arrieta-Montiel, M.; Christensen, A.C.; MacKenzie, S.A. Plant Mitochondrial Recombination Surveillance Requires Unusual RecA and MutS Homologs. *Plant Cell* **2007**, *19*, 1251–1264. [CrossRef] [PubMed]

22. Khazi, F.R.; Edmondson, A.C.; Nielsen, B.L. An Arabidopsis homologue of bacterial RecA that complements an *E. coli* recA deletion is targeted to plant mitochondria. *Mol. Genet. Genom.* **2003**, *269*, 454–463. [CrossRef] [PubMed]

23. Wallet, C.; Le Ret, M.; Bergdoll, M.; Bichara, M.; Dietrich, A.; Gualberto, J.M. The RECG1 DNA Translocase Is a Key Factor in Recombination Surveillance, Repair, and Segregation of the Mitochondrial DNA in *Arabidopsis*. *Plant Cell* **2015**, *27*, 2907–2925. [CrossRef]

24. Kobayashi, Y.; Misumi, O.; Odahara, M.; Ishibashi, K.; Hirono, M.; Hidaka, K.; Endo, M.; Sugiyama, H.; Iwasaki, H.; Kuroiwa, T.; et al. Holliday junction resolvases mediate chloroplast nucleoid segregation. *Science* **2017**, *356*, 631–634. [CrossRef] [PubMed]

25. Odahara, M.; Kishita, Y.; Sekine, Y. MSH1 maintains organelle genome stability and genetically interacts with RECA and RECG in the moss *Physcomitrella patens*. *Plant J.* **2017**, *91*, 455–465. [CrossRef]

26. Xu, Y.Z.; Arrieta-Montiel, M.P.; Virdi, K.S.; De Paula, W.B.; Widhalm, J.R.; Basset, G.J.; Davila, J.I.; Elthon, T.E.; Elowsky, C.G.; Sato, S.J.; et al. MutS HOMOLOG1 is a nucleoid protein that alters mitochondrial and plastid properties and plant response to high light. *Plant Cell* **2011**, *23*, 3428–3441. [CrossRef]

27. Lusetti, S.L.; Cox, M.M. The Bacterial RecA Protein and the Recombinational DNA Repair of Stalled Replication Forks. *Annu. Rev. Biochem.* **2002**, *71*, 71–100. [CrossRef]

28. Majeran, W.; Friso, G.; Asakura, Y.; Qu, X.; Huang, M.; Ponnala, L.; Watkins, K.P.; Barkan, A.; van Wijk, K.J. Nucleoid-enriched proteomes in developing plastids and chloroplasts from maize leaves: A new conceptual framework for nucleoid functions. *Plant Physiol.* **2012**, *158*, 156–189. [CrossRef]

29. McGlynn, P.; Lloyd, R.G. Rescue of stalled replication forks by RecG: Simultaneous translocation on the leading and lagging strand templates supports an active DNA unwinding model of fork reversal and Holliday junction formation. *Proc. Natl. Acad. Sci. USA* **2001**, *98*, 8227–8234. [CrossRef]

30. Cárdenas, P.P.; Carrasco, B.; Soufo, C.D.; César, C.E.; Herr, K.; Kaufenstein, M.; Graumann, P.L.; Alonso, J.C. RecX Facilitates Homologous Recombination by Modulating RecA Activities. *PLoS Genet.* **2012**, *8*, e1003126. [CrossRef]

31. Rocha, E.P.C.; Cornet, E.; Michel, B. Comparative and Evolutionary Analysis of the Bacterial Homologous Recombination Systems. *PLoS Genet.* **2005**, *1*, e15. [CrossRef]

32. Abdelnoor, R.V.; Yule, R.; Elo, A.; Christensen, A.C.; Meyer-Gauen, G.; MacKenzie, S.A. Substoichiometric shifting in the plant mitochondrial genome is influenced by a gene homologous to MutS. *Proc. Natl. Acad. Sci. USA* **2003**, *100*, 5968–5973. [CrossRef] [PubMed]

33. Culligan, K.M. Evolutionary origin, diversification and specialization of eukaryotic MutS homolog mismatch repair proteins. *Nucleic Acids Res.* **2000**, *28*, 463–471. [CrossRef] [PubMed]

34. Martinez-Zapater, J.M.; Gil, P.; Capel, J.; Somerville, C.R. Mutations at the Arabidopsis CHM locus promote rearrangements of the mitochondrial genome. *Plant Cell* **1992**, *4*, 889–899. [PubMed]

35. Sakamoto, W.; Kondo, H.; Murata, M.; Motoyoshi, F. Altered mitochondrial gene expression in a maternal distorted leaf mutant of Arabidopsis induced by chloroplast mutator. *Plant Cell* **1996**, *8*, 1377–1390.

36. Miller-Messmer, M.; Kühn, K.; Bichara, M.; Le Ret, M.; Imbault, P.; Gualberto, J.M. RecA-Dependent DNA Repair Results in Increased Heteroplasmy of the Arabidopsis Mitochondrial Genome. *Plant Physiol.* **2012**, *159*, 211–226. [CrossRef]

37. Davila, J.I.; Arrieta-Montiel, M.P.; Wamboldt, Y.; Cao, J.; Hagmann, J.; Shedge, V.; Xu, Y.Z.; Weigel, D.; Mackenzie, S.A. Double-strand break repair processes drive evolution of the mitochondrial genome in Arabidopsis. *BMC Biol.* **2011**, *9*, 64. [CrossRef]

38. Fukui, K.; Harada, A.; Wakamatsu, T.; Minobe, A.; Ohshita, K.; Ashiuchi, M.; Yano, T. The GIY-YIG endonuclease domain of Arabidopsis MutS homolog 1 specifically binds to branched DNA structures. *FEBS Lett.* **2018**, *592*, 4066–4077. [CrossRef]

39. Rowan, B.A.; Oldenburg, D.J.; Bendich, A.J. RecA maintains the integrity of chloroplast DNA molecules in Arabidopsis. *J. Exp. Bot.* **2010**, *61*, 2575–2588. [CrossRef]

40. Maréchal, A.; Parent, J.S.; Véronneau-Lafortune, F.; Joyeux, A.; Lang, B.F.; Brisson, N. Whirly proteins maintain plastid genome stability in Arabidopsis. *Proc. Natl. Acad. Sci. USA* **2009**, *106*, 14693–14698. [CrossRef]

41. Zampini, É.; Lepage, É.; Tremblay-Belzile, S.; Truche, S.; Brisson, N. Organelle DNA rearrangement mapping reveals U-turn-like inversions as a major source of genomic instability in Arabidopsis and humans. *Genome Res.* **2015**, *25*, 645–654. [CrossRef] [PubMed]

42. Odahara, M.; Kobayashi, Y.; Shikanai, T.; Nishimura, Y. Dynamic Interplay between Nucleoid Segregation and Genome Integrity in Chlamydomonas Chloroplasts. *Plant Physiol.* **2016**, *172*, 2337–2346. [CrossRef] [PubMed]

43. Maul, J.E.; Lilly, J.W.; Cui, L.; Depamphilis, C.W.; Miller, W.; Harris, E.H.; Stern, D.B. The *Chlamydomonas reinhardtii* plastid chromosome: Islands of genes in a sea of repeats. *Plant Cell* **2002**, *14*, 2659–2679. [CrossRef] [PubMed]

44. Sato, S. Complete Structure of the Chloroplast Genome of *Arabidopsis thaliana*. *DNA Res.* **1999**, *6*, 283–290. [CrossRef] [PubMed]

45. Vahrenholz, C.; Riemen, G.; Pratje, E.; Dujon, B.; Michaelis, G. Mitochondrial DNA of *Chlamydomonas reinhardtii*: The structure of the ends of the linear 15.8-kb genome suggests mechanisms for DNA replication. *Curr. Genet.* **1993**, *24*, 241–247. [CrossRef] [PubMed]

46. Sloan, D.B.; Wu, Z.; Sharbrough, J. Correction of Persistent Errors in *Arabidopsis* Reference Mitochondrial Genomes. *Plant Cell* **2018**, *30*, 525–527. [CrossRef]

47. Kurtz, S. REPuter: The manifold applications of repeat analysis on a genomic scale. *Nucleic Acids Res.* **2001**, *29*, 4633–4642. [CrossRef]

Article

Mitochondrial DNA Repair in an *Arabidopsis thaliana* Uracil N-Glycosylase Mutant

Emily Wynn [1,2], Emma Purfeerst [1,3] and Alan Christensen [1,*]

[1] School of Biological Sciences, University of Nebraska-Lincoln, Lincoln, NE 68588, USA;
    emilywynn6@gmail.com (E.W.); Emma.purfeerst@gmail.com (E.P.)
[2] United States Department of Agriculture, Agricultural Research Service, U.S. Meat Animal Research Center,
    Clay Center, NE 68933, USA
[3] Athletics Department, Bethany Lutheran College, Mankato, MN 56001, USA
* Correspondence: achristensen2@unl.edu; Tel.: +1-402-472-0681; Fax: +1-402-472-8722

Received: 19 December 2019; Accepted: 16 February 2020; Published: 18 February 2020

Abstract: Substitution rates in plant mitochondrial genes are extremely low, indicating strong selective pressure as well as efficient repair. Plant mitochondria possess base excision repair pathways; however, many repair pathways such as nucleotide excision repair and mismatch repair appear to be absent. In the absence of these pathways, many DNA lesions must be repaired by a different mechanism. To test the hypothesis that double-strand break repair (DSBR) is that mechanism, we maintained independent self-crossing lineages of plants deficient in uracil-N-glycosylase (UNG) for 11 generations to determine the repair outcomes when that pathway is missing. Surprisingly, no single nucleotide polymorphisms (SNPs) were fixed in any line in generation 11. The pattern of heteroplasmic SNPs was also unaltered through 11 generations. When the rate of cytosine deamination was increased by mitochondrial expression of the cytosine deaminase APOBEC3G, there was an increase in heteroplasmic SNPs but only in mature leaves. Clearly, DNA maintenance in reproductive meristem mitochondria is very effective in the absence of UNG while mitochondrial genomes in differentiated tissue are maintained through a different mechanism or not at all. Several genes involved in DSBR are upregulated in the absence of UNG, indicating that double-strand break repair is a general system of repair in plant mitochondria. It is important to note that the developmental stage of tissues is critically important for these types of experiments.

Keywords: mitochondria; DNA repair; double-strand break repair; uracil-N-glycosylase

1. Introduction

Plant mitochondrial genomes have very low base substitution rates but expand and rearrange rapidly [1–5]. The low substitution rate and the high rearrangement rate of plant mitochondria can be explained by selection and the specific DNA damage-repair mechanisms available. These mechanisms can also account for the genome expansions often found in land plant mitochondria [6]. The low nonsynonymous substitution rates in protein coding genes indicate that selective pressure to maintain the genes is high, and the low synonymous substitution rates indicate that the DNA-repair mechanisms are very accurate [7,8]. Despite the low mutation rate of mitochondrial genes over evolutionary time, mitochondrial genomes in mature cells accumulate DNA damage that is not repaired [9]. This indicates that there are fundamental differences between DNA maintenance in genomes meant to be passed on to the next generation and genomes that are not. In meristematic cells, mitochondria fuse together to form a large mitochondrion [10]. This fusion brings mitochondrial genomes together for genome replication but also ensures that there is a homologous template available for DNA repair. These meristematic cells eventually produce the reproductive tissue of a plant; from embryogenesis to egg cell production, the mitochondrial genomes inherited from parents and passed down to offspring will have homologous templates available to them [11].

Much less is known about the multiple pathways of DNA repair in plant mitochondria than in other systems, such as the nucleus. So far, there is no evidence of nucleotide excision repair (NER) or mismatch repair (MMR) in plant mitochondria [12,13]. It has been hypothesized that, in plant mitochondria, the types of DNA damage that are usually repaired through NER and MMR are repaired through double-strand break repair (DSBR) [14,15]. Plant mitochondria do have the nuclear-encoded base excision repair (BER) pathway enzyme Uracil DNA glycosylase (UNG) [12]. UNG is an enzyme that can recognize and bind to uracil in DNA and that can begin the process of base excision repair by enzymatically excising uracil (U) from single-stranded or double-stranded DNA [16]. Uracil can appear in a DNA strand due to the spontaneous deamination of cytosine or by the misincorporation of dUTP during replication [17]. Unrepaired uracil in DNA can lead to G-C to A-T transitions within the genome.

In light of the apparent absences of NER and MMR in plant mitochondria, it is possible that many lesions, including mismatches, are repaired by creating double-strand breaks and by using a template to repair both strands. Our hypothesis is that DSBR accounts for most of the repair in meristematic plant mitochondria and that both error-prone and accurate subtypes of DSBR lead to the observed patterns of genome evolution [18]. One way of testing this is to eliminate the pathway of uracil base excision repair and to ask if the G-U mispairs that occur by spontaneous deamination are repaired and, if so, are instead repaired by DSBR. In this work, we examine an *Arabidopsis thaliana UNG* knockout line and investigate the effects on the mitochondrial genome over many generations. To disrupt the genome further, we express the cytidine deaminase APOBEC3G in the *Arabidopsis* mitochondria (MTP-A3G) to increase the rate of cytosine deamination and to accelerate DNA damage.

One of the hallmarks of DSBR in plant mitochondria is the effect on the non-tandem repeats that exist in virtually all plant mitochondria [19]. The *Arabidopsis thaliana* mitochondrial genome contains two pairs of very large repeats (4.2 and 6.6 kb) that commonly undergo recombination [20–22], producing multiple isoforms of the genome. The mitochondrial genome also contains many non-tandem repeats between 50 and 1000 base pairs [19,22–24]. In wild type plants, these repeats recombine at very low rates, but they have been shown to recombine with ectopic repeat copies at higher rates in several mutants in DSBR-related genes, such as *msh1* and *reca3* [25–27]. Thus, genome dynamics around non-tandem repeats can be an indicator of increased DSBs. In this work, we show that a loss of uracil base excision repair leads to alterations in repeat dynamics, allowing us to observe an increase in genome abandonment in older leaves.

Numerous proteins known to be involved in the processing of plant mitochondrial DSBs have been characterized. Plants lacking the activity of mitochondrially targeted *recA* homologs have been shown to be deficient in DSBR [26,28]. In addition, it has been hypothesized that the plant MSH1 protein may be involved in binding to DNA lesions and in initiating DSBs [14,15]. The MSH1 protein contains a mismatch binding domain fused to a GIY-YIG type endonuclease domain which may be able to make DSBs [29,30], although an in vitro assay with a C-terminal fragment of the protein had no detectable endonuclease activity [31]. In this work, we provide evidence that, in the absence of mitochondrial UNG activity, several genes involved in DSBR, including *MSH1*, are transcriptionally upregulated, providing a possible explanation for the increased DSBR. We also provide additional evidence to support the hypothesis that mitochondrial DNA maintenance is abandoned in non-meristematic tissue [32], calling attention to the need to closely control for age and developmental state in experiments involving the mitochondrial genome.

2. Results

2.1. Lack of UNG Activity in Mutants

It has previously been reported that cell extracts of the *Arabidopsis thaliana* UNG T-DNA insertion strain used in this experiment, GK-440E07 (ABRC seed stock CS308282), show no uracil glycosylase activity [12]. To increase the rate of cytosine deamination in the mitochondrial genome and to show that effects of the UNG knockout on mitochondrial mutation rates could be detected, the catalytic domain of the human APOBEC3G–CTD 2K3A cytidine deaminase (A3G) [33] was expressed under the control of the ubiquitin-10 promoter [34] in both wild-type and UNG *Arabidopsis thaliana* lines and targeted the mitochondria by an amino-terminal fusion of the 62 amino acid mitochondrial targeting peptide (MTP) from the alternative oxidase 1A protein. Fluorescence microscopy of *Arabidopsis thaliana* expressing an MTP-A3G-GFP fusion shows that the MTP-A3G construct is expressed and targeted the mitochondria (Figure S1).

We expected that, in the absence of UNG, there would be an increase in G-C to A-T substitution mutations. To test this prediction, we sequenced a wild-type Arabidopsis plant (Col-0), a wild-type Arabidopsis plant expressing the MTP-A3G construct (Col-0 MTP-A3G), and a UNG plant expressing the MTP-A3G construct (UNG MTP-A3G) using an Illumina Hi-Seq4000 system. Mitochondrial sequence reads from these plants were aligned to the Columbia-0 reference genome (modified as described in the Materials and Methods section) using BWA-MEM [35], and single nucleotide polymorphisms were identified using VarDict [36]. VarDict was chosen due to its high sensitivity and accuracy compared with other low-frequency variant callers when analyzing Illumina HiSeq data [37].

There were no SNPs that reached fixation (an allele frequency of 1) in any plant. Mitochondrial genomes are not diploid; each cell can have many copies of the mitochondrial genome. Therefore, it is possible that an individual plant could accumulate low-frequency mutations in some of the mitochondrial genomes in the cell. VarDict was used to detect heteroplasmic SNPs at allele frequencies as low as 0.01. VarDict's sensitivity in calling low-frequency SNPs scales with depth of coverage and quality of the sample, so it is not possible to directly compare heteroplasmic mutation rates in samples with different depths of coverage. However, because the activity of the UNG protein is specific to uracil, the absence of the UNG protein should not have any effect on mutation rates other than G-C to A-T transitions. Comparing the numbers of G-C to A-T transitions to all other substitutions should reveal if the rate of mutations that can be repaired by UNG is elevated compared to the background rate. If the UNG MTP-A3G line is accumulating G-C to A-T transitions at a faster rate than the Col-0 MTP-A3G line, we would expect to see an increased ratio of G-C to A-T transitions compared to other mutation types. Complicating the analysis, significant portions of the *A. thaliana* mitochondrial genome have been duplicated in the nucleus, forming regions called NuMTs, an abbreviation of Nuclear Mitochondrial DNA [38–40]. Mutations in the NuMTs might appear to be low-frequency SNPs in the mitochondrial genome, confounding the results. However, these mutations are likely to be shared in the common nuclear background of all our lines. To avoid attributing SNPs in NuMTs to the mitochondrial genome, only those SNPs unique to individual plant lines were used in this comparison. In addition, many of the shared SNPs were flanked by a number of paired-end reads with one end in the mitochondrial genome and the other in the nuclear genome, additional evidence that they are NuMTs. The Col-0 plant had a heteroplasmic GC-AT/total SNPs ratio of 0, the Col-0 MTP-A3G plant had a heteroplasmic GC-AT/total SNPs ratio of 0.47, while the UNG MTP-A3G plant had a heteroplasmic GC-AT/total SNPs ratio of 0.92 (Table 1). Therefore, when the rate of cytosine deamination is increased by the activity of APOBEC3G, Arabidopsis plants accumulate GC-AT SNPs and our computational pipeline is able to detect this increase.

Table 1. Heteroplasmic mitochondrial single nucleotide polymorphisms (SNPs) in Col-0 wild-type, generation 10 *uracil DNA glycosylase (UNG)* mutant lines, Col-0 MTP-A3G, and *UNG* MTP-A3G: SNPs were called using VarDict as described in the Methods section. SNP counts are shown for the entire mitochondrial genome. For the full spectrum of SNP types, including allele frequencies, see Supplementary File 2.

Sample	A-C	A-G	A-T	C-A	C-G	C-T	G-A	G-C	G-T	T-A	T-C	T-G	Total	GC-AT/Total Ratio
Col-0	0	17	1	0	0	0	0	0	0	0	11	0	29	0
*UNG*10 115	0	7	0	0	0	0	0	0	1	0	7	0	15	0
*UNG*10 159	0	15	0	0	0	0	0	0	0	0	21	0	36	0
*UNG*10 163	1	305	10	3	1	3	7	0	1	21	281	0	633	0.016
*UNG*10 176	0	97	3	0	0	0	1	0	2	4	76	0	183	0.0055
*UNG*10 198	0	9	0	0	0	1	0	0	0	1	6	0	17	0.059
*UNG*10 201	0	12	1	0	0	0	1	0	0	1	15	0	30	0.033
*UNG*10 203	0	9	1	0	0	0	0	0	0	1	5	0	16	0
Col-0 MTP A3G	0	6	0	0	0	3	4	0	0	0	2	0	15	0.47
UNG MTP A3G	0	0	0	0	0	5	7	0	0	0	1	0	13	0.92

2.2. Mutation Accumulation in the Absence of UNG

To determine the effects of the *UNG* knockout across multiple generations, we performed a mutation accumulation study [41]. We chose 23 different *UNG* homozygous plants derived from one hemizygous parent. These 23 plants were designated as generation 1 *UNG* and were allowed to self-cross. The next generation was derived by single-seed descents from each line, and this was repeated until generation 10 *UNG* plants were obtained. Leaf tissue and progeny seeds from each line were kept at each generation.

The leaf tissue from generation 10 of the *UNG* mutation accumulation lines and a wild-type Col-0 were sequenced and analyzed with VarDict as described above. Similar to the MTP-A3G plants, there were no SNPs in any of our *UNG* mutation accumulation lines that had reached fixation (an allele frequency of one). In contrast, there was no relative increase in the ratios of GC-AT/total SNPs between the *UNG* lines and Col-0 (see Table 1). Because detection of low-frequency SNPs depends on read depth, we only report the 7 *UNG* samples with an average mitochondrial read depth above 125× for this comparison. In the absence of a functional UNG protein and under normal greenhouse physiological conditions, plant mitochondria do not accumulate cytosine deamination mutations at an increased rate.

2.3. Nuclear Mutation Accumulation

UNG is the only uracil-N-glycosylase in *Arabidopsis thaliana* and may be active in the nucleus as well as the mitochondria [12]. To test for nuclear mutations due to the absence of UNG, sequences were aligned to the Columbia-0 reference genome using BWA-MEM and single nucleotide polymorphisms were identified using Bcftools Call [42]. The *UNG* mutation accumulation lines do not have an elevated G-C to A-T mutation rate compared to wild-type (Table 2).

2.4. Alternative Repair Pathway Genes

Because the *UNG* mutants show increased double-strand break repair but not an increase of G-C to A-T transition mutations, we infer that the inevitable appearance of uracil in the DNA is repaired via conversion of a G-U pair to a double-strand break and efficiently repaired by the DSBR pathway. If this is true, genes involved in the DSBR processes of breakage, homology surveillance, and strand invasion in mitochondria will be upregulated in *UNG* mutants. To test this hypothesis, we assayed transcript levels of several candidate genes known to be involved in DSBR [13,23,25–28,43–46] in *UNG* lines compared to wild-type using RT-PCR. *MSH1* and *RECA2* were significantly upregulated in *UNG* lines (*MSH1*: 5.60-fold increase, unpaired T-test $p < 0.05$. *RECA2*: 3.19-fold increase, unpaired T-test $p < 0.05$; see Figure 1). The single-strand binding protein gene *OSB1* was also measurably upregulated in *UNG* lines (3.07-fold increase, unpaired T-test $p = 0.053$). *RECA3*, *SSB*, and *WHY2* showed no significant differential expression compared to wild-type (unpaired T-test $p > 0.05$).

Table 2. Nuclear SNPs in Col-0 wild-type, *UNG* mutant lines, Col-0 MTP-A3G, and *UNG* MTP-A3G: SNPs were called using Bcftools Call as described in the Methods section. SNP counts are for each chromosome, excluding chromosome 2. For individual data on each chromosome, see Supplementary File 2.

Sample	GC-AT SNPs	Total SNPs	GC-AT/Total Ratio
Col-0	0	0	0
*UNG*10 115	21	69	0.30
*UNG*10 159	27	79	0.34
*UNG*10 163	10	43	0.23
*UNG*10 176	28	74	0.38
*UNG*10 198	23	73	0.32
*UNG*10 201	53	131	0.40
*UNG*10 203	35	97	0.36
Col-0 MTP-A3G	0	0	0
UNG MTP-A3G	44	154	0.29
Col-0 1 Young	31	81	0.38
Col-0 2 Young	44	101	0.44
*UNG*11 163 1 Young	31	111	0.28
*UNG*11 163 2 Young	59	141	0.42
*UNG*11 176 1 Young	32	149	0.21
*UNG*11 176 2 Young	48	126	0.38
*UNG*11 198 1 Young	60	236	0.25
*UNG*11 198 2 Young	99	311	0.32
Col-0 MTP-A3G 1 Young	17	53	0.32
Col-0 MTP-A3G 2 Young	25	58	0.43
UNG MTP-A3G 1 Young	130	453	0.29
UNG MTP-A3G 2 Young	589	2711	0.22
Col-0 1 Mature	17	47	0.36
Col-0 2 Mature	15	43	0.35
*UNG*11 163 1 Mature	27	113	0.24
*UNG*11 163 2 Mature	23	77	0.30
*UNG*11 176 1 Mature	31	99	0.31
*UNG*11 176 2 Mature	27	93	0.29
*UNG*11 198 1 Mature	41	157	0.26
*UNG*11 198 2 Mature	66	195	0.34
Col-0 MTP-A3G 1 Mature	15	45	0.33
Col-0 MTP-A3G 2 Mature	13	37	0.35
UNG MTP-A3G 1 Mature	106	339	0.32
UNG MTP-A3G 2 Mature	175	684	0.26

2.5. Increased Mitochondrial Genome Abandonment

If most DNA damage in plant mitochondria is repaired by double-strand break repair (DSBR), supplemented by base excision repair [12], then in the absence of the Uracil-N-glycosylase (UNG) pathway, we predict an increase in DSBR. To find evidence of this, we used quantitative PCR (qPCR) to assay crossing over between identical non-tandem repeats because changes in the dynamics around these repeats is indicative of changes in DNA processing at double-strand breaks [26,27,46]. Different combinations of primers in the unique sequences flanking the repeats allow us to determine the relative copy numbers of parental-type repeats and low-frequency recombinants (Figure 2a). The mitochondrial genes *cox2* and *rrn18* were used to standardize relative amplification between lines. We and others [24,46] have found that some of the non-tandem repeats are well suited for qPCR analysis and are sensitive indicators of ectopic recombination, increasing in repair-defective mutants. We analyzed the three repeats known as repeats B, D, and L [23] in both young leaves and mature leaves. In young leaves, there is no significant difference in the amounts of parental or recombinant forms between *UNG* lines and Col-0 (Figure 2b). In mature leaves, all three repeats show significant reductions in the parental 2/2 form while repeat B also shows a reduction in the parental 1/1 form (unpaired t-test $p < 0.05$; Figure 2c). There is a difference in genome dynamics around non-tandem repeats in young leaves compared to old leaves, indicating a difference in the way these genomes are maintained.

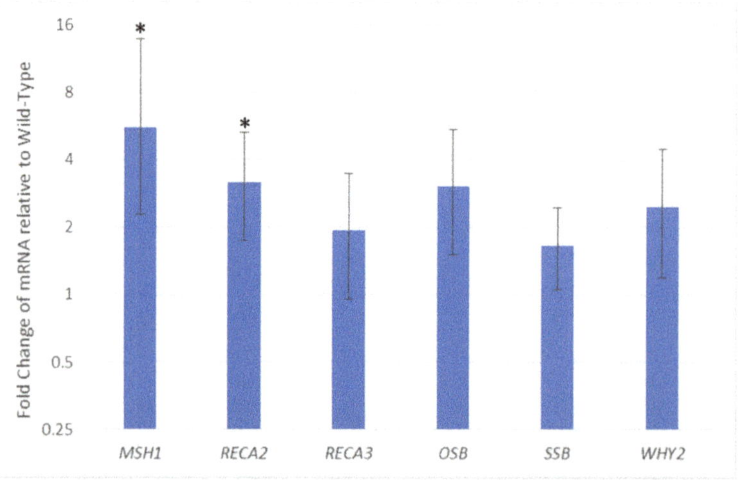

Figure 1. Quantitative RT-PCR assays of enzymes involved in double-strand break repair (DSBR) in *UNG* lines relative to wild-type: Fold change in transcript level is shown on the Y-axis. Error bars are standard deviation of three biological replicates. *MSH1* and *RECA2* are significantly transcriptionally upregulated in *UNG* lines relative to wild-type (5.60-fold increase and 3.19-fold increase, respectively. Unpaired, 2-tailed student's t-test, * indicates $p < 0.05$). *OSB1* is nearly significantly upregulated in *UNG* lines relative to wild-type (3.07-fold increase. Unpaired t-test $p = 0.053$).

2.6. Transmission of SNPs Across Generations

To determine if any heteroplasmic SNPs are passed on to the next generation, two progenies of each of the wild-type, *UNG*, MTP-A3G, and *UNG* MTP-A3G plants that were sequenced above were planted. Leaves were collected from each plant when it was 17 days old (young leaf) and again when it was 36 days old (mature leaf). Both the young and mature leaves of each plant were sequenced and analyzed as described above. Only 1 heteroplasmic SNP could be traced from a parent plant to both progeny, and 7 heteroplasmic SNPs could be traced from a parent plant to one progeny (Supplementary File 2). Interestingly, 117 heteroplasmic SNPs were detected in both offspring but not the parent plant. It is possible that heteroplasmic mutations that occur in reproductive tissue after the parental tissue had been collected could be passed on to the progeny. However, only 3 of these heteroplasmic SNPs are found in the mature tissue of both progeny, indicating that, even if a heteroplasmic SNP is passed on to a future generation, it is likely to be removed from the mitochondrial population before reproduction by genetic drift or gene conversion. In fact, of the 2792 heteroplasmic SNPs that were detected in young tissue across all samples, only 4 were detected in the mature tissue of the same plant. The overwhelming majority of heteroplasmic SNPs arose in mitochondria in non-meristematic differentiated tissue.

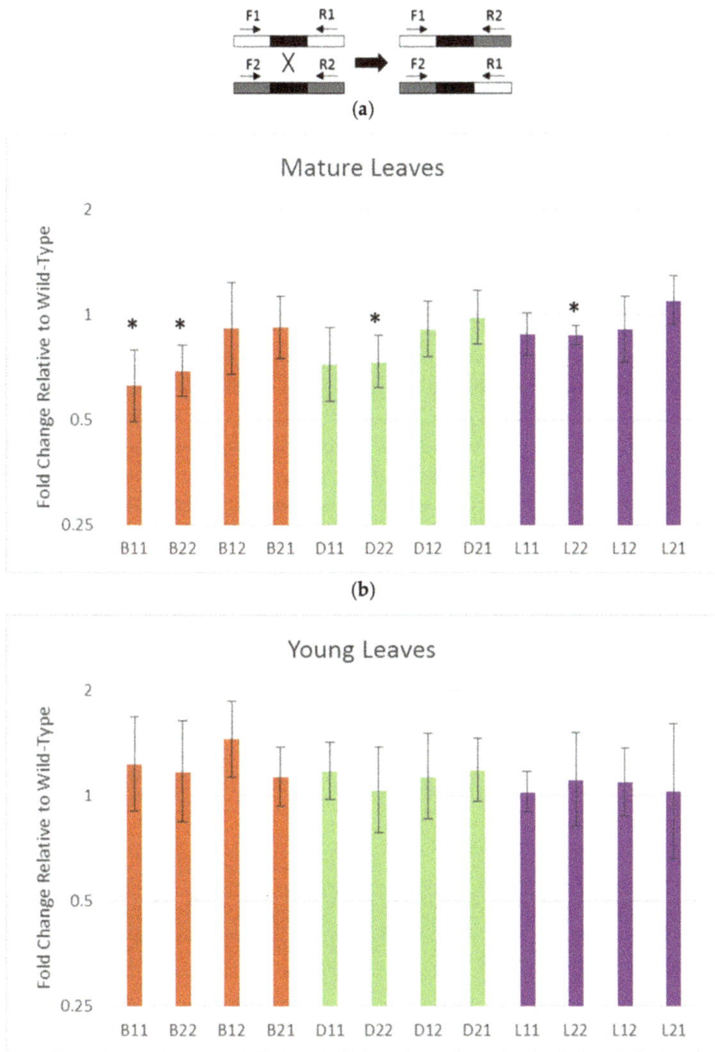

Figure 2. qPCR analysis of intermediate repeat recombination in *UNG* lines compared to wild-type: Recombination at intermediate repeats is an indicator of increased double-strand breaks in plant mitochondrial genomes. (**a**) Primer scheme for detecting parental and recombinant repeats: Using different combinations of primers that anneal to the unique sequence flanking the repeats, either parental type (1/1 and 2/2) or recombinant type (1/2 and 2/1) repeats can be amplified. (**b**) Fold change of intermediate repeats in young leaves of *UNG* lines relative to wild-type: Error bars are standard deviation of three biological replicates. (**c**) Fold change of intermediate repeats in mature leaves of *UNG* lines relative to wild-type: Error bars are standard deviation of three biological replicates. B1/1, B2/2, D2/2, and L2/2 show significant reduction in copy number (unpaired, 2-tailed student's t-test, * indicates $p < 0.05$).

2.7. SNP Accumulation in Young vs. Mature Leaves

To confirm that the effects of the UNG knockout and the expression of APOBEC3G are consistent, the progenies of the wild-type, *UNG*, MTP-A3G, and *UNG* MTP-A3G plants were analyzed and the ratio of heteroplasmic GC-AT to total heteroplasmic SNPs was compared as described above. In mature leaves, the results were similar to the previous generation: both the *UNG* MTP-A3G and Col-0 MTP-A3G samples had increased GC-AT SNPs compared to the *UNG* and Col-0. Interestingly, in young leaves, neither the *UNG* MTP-A3G nor the Col-0 MTP-A3G samples had increased GC-AT SNPs (See Table 3). This indicates that the processes of mitochondrial genome maintenance are more efficient at repairing DNA damage in young leaves.

2.8. Quality Control of DNA Library Preparation

A common source of error when calling low-frequency SNPs is oxidative damage during library preparation [47]. This oxidative damage affects guanines, and the effects of this damage can be measured by comparing the ratio of G to T mutations between the R1 paired-end read and the R2 paired-end read. A Global Imbalance Value (GIV) above 1.5 indicates DNA damage during library preparation, while a GIV below 1.5 indicates little damage during library preparation. None of the samples used in this study had a GIV above 1.5 (see Supplementary File 2), indicating that DNA damage during library prep is not a significant source of false SNP calls.

Table 3. Heteroplasmic mitochondrial SNPs in the next generation of Col-0 wild-type, *UNG mutant lines*, *Col-0 MTP-A3G*, and *UNG MTP-A3G:* SNPs were called using VarDict as described in the Methods section. SNP counts are shown for the entire mitochondrial genome. For the full spectrum of SNP types, including allele frequencies, see Supplementary File 2.

Sample	A-C	A-G	A-T	C-A	C-G	C-T	G-A	G-C	G-T	T-A	T-C	T-G	Total	GC-AT/Total Ratio
Col-0 1 Young	0	224	3	0	0	4	6	0	0	3	227	0	467	0.0214
Col-0 2 Young	0	45	1	0	0	2	0	0	0	2	33	1	84	0.0238
*UNG*11 163 1 Young	0	205	4	0	0	7	7	0	0	4	216	0	443	0.0316
*UNG*11 163 2 Young	1	27	2	0	0	0	1	0	0	0	47	0	78	0.0128
*UNG*11 176 1 Young	0	65	0	0	0	0	2	0	0	3	79	0	149	0.0134
*UNG*11 176 2 Young	0	194	0	0	0	2	5	0	0	7	171	0	379	0.0185
*UNG*11 198 1 Young	1	62	2	0	0	2	1	0	0	1	59	0	128	0.0234
*UNG*11 198 2 Young	0	63	2	0	0	0	1	0	0	4	68	0	138	0.00725
Col-0 MTP A3G 1 Young	0	151	5	2	0	11	15	0	1	2	158	0	345	0.0754
Col-0 MTP A3G 2 Young	0	154	2	0	0	7	3	0	1	6	136	0	309	0.0324
UNG MTP A3G 1 Young	0	84	0	0	0	3	4	0	0	2	82	0	175	0.04
UNG MTP A3G 2 Young	0	49	2	0	0	2	1	0	0	1	46	2	103	0.0291
Col-0 1 Mature	0	50	0	0	0	3	3	0	0	1	54	0	111	0.0541
Col-0 2 Mature	0	65	2	1	0	2	1	0	0	4	60	0	135	0.0222
*UNG*11 163 1 Mature	0	1	0	0	0	0	0	0	0	0	3	0	4	0
*UNG*11 163 2 Mature	0	2	0	0	0	0	0	0	0	0	6	0	8	0
*UNG*11 176 1 Mature	0	35	2	0	0	2	1	0	0	2	50	1	93	0.0323
*UNG*11 176 2 Mature	0	50	0	0	0	1	2	0	0	0	57	0	110	0.0273
*UNG*11 198 1 Mature	0	42	1	0	0	0	2	0	0	1	42	0	88	0.0227
*UNG*11 198 2 Mature	0	29	2	0	0	1	0	0	0	0	33	0	65	0.0154
Col-0 MTP A3G 1 Mature	0	4	0	0	0	0	1	0	0	0	3	0	8	0.125
Col-0 MTP A3G 2 Mature	0	14	0	0	0	9	12	0	0	0	10	0	45	0.467
UNG MTP A3G 1 Mature	0	13	0	0	0	33	41	0	0	0	13	0	100	0.74
UNG MTP A3G 2 Mature	0	78	1	0	0	25	36	0	1	0	69	0	210	0.290

3. Discussion

In mitochondria as well as in the nucleus and chloroplast, cytosine is subject to deamination to uracil. This could potentially lead to transition mutations and is dealt with by a specialized base excision repair pathway. The first step in this pathway is hydrolysis of the glycosidic bond by the enzyme Uracil-N-glycosylase (UNG), leaving behind an abasic site [16]. An AP (apurinic) endonuclease can then cut the DNA backbone, producing a 3′ OH and a 5′ dRP (5′-deoxyribose-5-phosphate). Both DNA polymerases found in *A. thaliana* mitochondria, POL1A and POL1B, exhibit 5′-dRP lyase activity, allowing them to remove the 5′ dRP and to polymerize a new nucleotide replacing the uracil [48]. In the absence of functional UNG protein, cytosine will still be deaminated in plant mitochondrial genomes, so efficient removal of uracil must be through a different repair mechanism, most likely DSBR [14,15]. We have found that, in *UNG* mutant lines, there is an increase in the expression of genes known to be involved in DSBR and significant changes in the relative abundance of parental and recombinant forms of intermediate repeats, consistent with this hypothesis.

We have shown that, when cytosine deamination is increased by the expression of the APOBEC3G cytidine deaminase in plant mitochondria, *UNG* lines accumulate more G-C to A-T transitions in mature leaves than does wild-type. Surprisingly, we have also found that, under normal cellular conditions, without the added deamination activity of APOBEC3G, *UNG* lines do not accumulate G-C to A-T transition mutations at a higher rate than wild-type. This finding is particularly surprising given the presumed bottlenecking of mitochondrial genomes during female gametogenesis and given the deliberate bottleneck in the experimental design of single-seed descent for 11 generations. This finding supports the hypothesis that plant mitochondria have a very efficient alternative damage surveillance system that can prevent G-C to A-T transitions from becoming fixed in the meristematic mitochondrial population. The use of the cytidine deaminase allows us to specifically alter the mutation rate, which helps us disentangle mutation from repair, selection and drift—common complications in mutation accumulation experiments [49].

The angiosperm MSH1 protein consists of a DNA mismatch-binding domain fused to a double-stranded DNA endonuclease domain [1,21] Although mainly characterized for its role in recombination surveillance [36], MSH1 is a good candidate for a protein that may be able to recognize and bind to various DNA lesions and to make DSBs near the site of the lesion, thus funneling these types of damage into the DSBR pathway. With many mitochondria and many mitochondrial genomes in each cell, there are numerous available templates for accurate repair of DSBs through homologous recombination, making this a plausible mechanism of genome maintenance. Here, we show that, in *UNG* lines, *MSH1* is transcriptionally upregulated more than 5-fold compared to wild-type. This is consistent with the hypothesis that MSH1 initiates repair in plant mitochondria by creating a double-strand break at G-U pairs and possibly other mismatches and damaged bases.

Several other proteins involved in processing plant mitochondrial DSBs have been characterized. The RECA homologs RECA2 and RECA3 are homology search and strand invasion proteins [26–28,45,50–52]. The two mitochondrial RECAs share much sequence similarity; however, RECA2 is dual targeted to both the mitochondria and the plastids, while RECA3 is found only in the mitochondria [26,27]. RECA3 also lacks a C-terminal motif present on RECA2 and most other homologs. This motif has been shown to modulate the ability of RECA proteins to displace competing ssDNA binding proteins in *E. coli* [53]. Arabidopsis *reca2* mutants are seedling lethal, and both *reca2* and *reca3* lines show increased ectopic recombination at intermediate repeats [26]. Arabidopsis RECA2 has functional properties that RECA3 cannot perform, such as complementing a bacterial *recA* mutant during the repair of UV-C-induced DNA lesions [20]. Here, we show that, in *UNG* lines, *RECA2* is transcriptionally upregulated more than 3-fold compared to the wild-type. However, *RECA3* is not upregulated in *UNG* lines. Responding to MSH1-initiated DSBs may be one of the functions unique to RECA2. The increased expression of *RECA2* in the absence of a functional UNG protein is further evidence that uracil arising in DNA may be repaired through the mitochondrial DSBR pathway.

The ssDNA binding protein OSB1's transcript is upregulated over 3-fold. At a double-strand break, OSB1 competitively binds to ssDNA and recruits the RECA proteins to promote the repair of a double-strand break by a homologous template and to avoid the error-prone microhomology-mediated end-joining pathway [54].

We also tested the differential expression of other genes known to be involved in processing mitochondrial DSBs. The single-stranded binding protein genes *WHY2* and *SSB* were not found to be differentially expressed at the transcript level compared to wild-type. The presence of different ssDNA binding proteins influences which pathway of DSBR a break is repaired by [54]. Increased amounts of WHY2 and SSB may not be needed for accurate repair of induced DSBs in the *UNG* lines.

At intermediate repeats, the maintenance of the mitochondrial genome is different between wild-type, *UNG* mutants, and DSBR mutants. In *msh1* lines, there is an increase in repeat recombination likely due to relaxed homology surveillance in the absence of the MSH1 protein [27]. In mutant lines of ssDNA binding proteins involved in DSBR, such as *recA2*, *recA3*, and *osb1* [26,55], there is an increase in repeat recombination due to differences in the way DNA ends are handled in the absence of these ssDNA binding proteins. In young leaves, there is no significant difference in recombination at intermediate repeats between *UNG* lines and wild-type, while in mature leaves, *UNG* lines show a reduction in parental type repeats compared to wild-type. In *UNG* lines, the mitochondrial recombination machinery is still intact, so any differences in genome dynamics at intermediate repeats are not due to differences in processing the DSBs; instead, this could indicate that there is an increase in double-strand breaks and an increase in attempted DSBR by break-induced replication at intermediate repeats or that this could be an indication of degradation of mtDNA as differentiated tissue ages.

Plant mitochondrial genomes likely replicate by recombination-dependent replication (RDR) [56]. Most organellar genome replication occurs in meristematic tissue, where mitochondria fuse together to form a large, reticulate mitochondrion [10]. This mitochondrial fusion provides a means to homogenize mtDNA by gene conversion and to repair lesions through homologous recombination [57]. Accurate repair of uracil by homologous recombination would not be expected to change repeat dynamics. As cells differentiate and age, organellar genomes degrade [32]. Organellar genomes in nonreproductive tissue can be "abandoned" rather than repaired, reducing the metabolic cost of DNA repair [32]. In a mature cell, an attempt to repair uracil in the mitochondrial genome could lead to degradation of the DNA and changes in repeat dynamics if a double-strand break is initiated without a homologous template available. There is a difference in mitochondrial DNA maintenance in mature cells compared to young cells, due to either a lack of DNA repair in mature mitochondria or a difference in DNA-repair mechanism.

To determine the outcomes of genomic uracil in the absence of a functional UNG protein, we sequenced the genomes of several *UNG* lines. No fixed mutations of any kind were found in *UNG* lines, even after 11 generations of self-crossing. Low-frequency heteroplasmic SNPs were found in both wild-type and *UNG* lines, but *UNG* lines showed no difference in the ratio of G-C to A-T transitions to other mutation types when compared to wild-type. When the rate of cytosine deamination was increased with the expression of the APOBEC3G deaminase, there was an increase in G-C to A-T transitions but only in mature leaves. This is consistent with the idea of abandonment and is evidence that, in mitochondrial genomes that have not been abandoned, there is an efficient and accurate system of nonspecific repair.

Clearly, plant mitochondria can repair uracil in DNA sufficiently to prevent mutation accumulation in the absence of the UNG protein. Why then has the BER pathway been conserved in plant mitochondria while NER and MMR have apparently been lost? DSBR may be able to protect the genome efficiently from mutations being inherited by the next generation (see Table 3). There may still be selection to maintain mitochondrial BER to reduce the rate of mitochondrial genome abandonment and degradation in aging tissues. Throughout the evolutionary history of *Arabidopsis thaliana* and into the present, wild growing plants are exposed to a range of growth conditions and stresses that experimental plants in a greenhouse avoid. The rate of spontaneous cytosine deamination increases with increasing temperature [58,59], so DSBR alone may not be able repair the extent of uracil found in DNA across the range of temperatures a wild plant would experience, providing the selective pressure to maintain a distinct BER pathway in plant mitochondria. If DSBR activity is reduced or lost as leaf tissue ages, there may also be a selective advantage to the plant of maintaining BER in mature leaves so they can continue to perform intermediary metabolism even as they age.

Here, we have provided evidence that, in the absence of a dedicated BER pathway, plants growing in greenhouse growth chamber conditions do not accumulate mitochondrial SNPs at an increased rate. Instead, DNA damage is accurately repaired by double-strand break repair, which also causes an increase in ectopic recombination at identical non-tandem repeats. It has recently been shown that mice lacking a different mitochondrial BER protein, oxoguanine glycosylase, also do not accumulate mitochondrial SNPs [60]. Here, we show that, in plants, base-excision repair by UNG is similarly unnecessary to prevent mitochondrial mutations in growth chamber conditions. The presence of the UNG pathway reduces ectopic recombination slightly and can successfully repair uracil in DNA even if the rate of cytosine deamination is increased. We have also found that, in mature leaves, uracil mutations do occur, further confirming the hypothesis that organellar genomes are abandoned in terminally differentiated tissues [32] and emphasizing the need for considering the tissue age and type when interpreting experimental results on DNA replication, repair, and recombination. Double-strand break repair and recombination are important mechanisms in the evolution of plant mitochondrial genomes, but many key enzymes and steps in the repair pathway are still unknown. Further identification and characterization of these missing steps is sure to provide additional insight into the unique evolutionary dynamics of plant mitochondrial genomes.

4. Materials and Methods

4.1. Plant Growth Conditions

Arabidopsis thaliana Columbia-0 (Col-0) seeds were obtained from Lehle Seeds (Round Rock, TX, USA). UNG (AT3G18630) T-DNA insertion hemizygous lines were obtained from the Arabidopsis Biological Resource Center, line number CS308282. Hemizygous T-DNA lines were self-crossed to obtain homozygous lines (Genotyping primers: wild-type 5′-TGTCAAAGTC CTGCAATTCTTCTCACA-3′ and 5′-TCGTGCCATATCTTGCAGACCACA-3′, and *UNG* 5′-ATA ATAACGCTGCGGACATCTACATTTT-3′ and 5′-ACTTGGAGAAGGTAAAGCAATTCA-3′). All plants were grown in walk-in growth chambers under a 16:8 light:dark schedule at 22 °C. Plants grown on agar were surface sterilized and grown on 1× Murashige and Skoog Basal Medium (MSA) with Gamborg's vitamins (Sigma, St. Louis, MO, USA) with 5 μg/mL Nystatin Dihydrate to prevent fungal contamination.

4.2. Vector Construction

The APOBEC3G gene [61] was synthesized by Life Technologies Gene Strings using *Arabidopsis thaliana*-preferred codons and including the 62 amino acid mitochondrial targeting peptide (MTP) from alternative oxidase on the N-terminus of the translated protein. The MTP-A3G construct was cloned into the vector pUB-DEST (NCBI:taxid1298537) driven by the ubiquitin (UBQ10) promoter and transformed into wild-type and *UNG Arabidopsis thaliana* plants by the *Agrobacterium* floral dip method [62]. To ensure proper mitochondrial targeting of the MTP-A3G construct, the construct was cloned into pK7FWG2 with a C-terminal GFP fusion [63]. *Arabidopsis thaliana* plants were again transformed by the *Agrobacterium* floral dip method, and mitochondrial fluorescence was confirmed with confocal fluorescence microscopy.

4.3. RT-PCR

RNA was extracted from young leaves of plants grown in parallel on MSA during *UNG* generation ten [64]. Reverse transcription using Bio-Rad iScript was performed, and the resulting cDNA was used as a template for qPCR to measure relative transcript amounts. Quantitative RT-PCR data was normalized using *UBQ10* and *GAPDH* as housekeeping gene controls. Reactions were performed in a Bio-Rad CFX96 thermocycler using 96-well plates and a reaction volume of 20 μL/well. SYBRGreen mastermix (Bio-Rad, Hercules, CA, USA) was used in all reactions. Three biological replicates from different *UNG* MA lines and three technical replicates were used for each amplification. Primers are listed in Table S1. The MIQE guidelines were followed [65], and primer efficiencies are listed in Table S2. The thermocycling program for all RT-qPCR was a ten-minute denaturing step at 95° followed by 45 cycles of 10 s at 95°, 15 s at 60°, and 13 s at 72°. Following amplification, melt curve analysis was done on all reactions to ensure target specificity. The melt curve program for all RT-qPCR was from 65°–95° at 0.5° increments for 5 s each.

4.4. Repeat Recombination qPCR

DNA was collected from young and mature leaves of Columbia-0 and generation ten *UNG* plants grown in parallel using the CTAB DNA extraction method [66]. qPCR was performed using primers from the flanking sequences of the intermediate repeats. Primers are listed in Table S1. Using different combinations of forward and reverse primers, either the parental or recombinant forms of the repeat can be selectively amplified (see Figure 2a). The mitochondrially encoded *cox2* and *rrn18* genes were used as standards for analysis. Reactions were performed in a Bio-Rad CFX96 thermocycler using 96-well plates with a reaction volume of 20μL/well. SYBRGreen mastermix (Bio-Rad) was used in all reactions. Three biological and three technical replicates were used for each reaction. The thermocycling program for all repeat recombination qPCR was a ten-minute denaturing step at 95° followed by 45 cycles of 10 s at 95°, 15 s at 60°, and a primer-specific amount of time at 72° (extension times for each primer pair can be found in Table S3). Following amplification, melt curve analysis was done on all reactions to ensure target specificity. The melt curve program for all qPCR was from 65°–95° at 0.5° increments for 5 s each.

4.5. DNA Sequencing

DNA extraction from frozen mature leaves of Columbia-0, generation 10 and *UNG*, and MTP-A3G plants and again from young and mature leaves of the progeny of these plants was done by a modification of the SPRI (Solid Phase Reversible Immobilization) magnetic beads method of Rowan et al. [67,68]. Genomic libraries for paired-end sequencing were prepared using a modification of the Nextera protocol [69] and modified for smaller volumes following Baym et al. [70]. Following treatment with the Nextera Tn5 transpososome, 14 cycles of amplification were done. Libraries were size-selected to be between 400 and 800 bp in length using SPRI beads [68]. Libraries were sequenced with 150 bp paired-end reads on an Illumina HiSeq 4000 by the Vincent J. Coates Genomics Sequencing Laboratory at UC Berkeley. The raw data files are deposited with the Sequence Read Archive at ncbi.nlm.nih.gov under BioProject number PRJNA492503.

Reads were aligned using BWA-MEM v0.7.12-r1039 [35]. The reference sequence used for alignment was a file containing the improved Columbia-0 mitochondrial genome (accession BK010421.1) [71] as well as the TAIR 10 *Arabidopsis thaliana* nuclear chromosomes and chloroplast genome sequences [72]. A large portion of the mitochondrial genome has been duplicated into chromosome 2 [40]. To prevent reads from mapping to both locations, this large NuMT region was deleted from chromosome 2. Using Samtools v1.3.1 [73], bam files were sorted for uniquely mapped reads for downstream analysis. MarkDuplicates from the Genome Analysis ToolKit (GATK) was used to remove duplicate reads due to PCR during library prep [74].

Organellar variants were called using VarDict [36]. To minimize the effects of sequencing errors and to reduce false positives, SNPs called by VarDict were filtered by the stringent quality parameters of Qmean $\geq$ 30, MQ $\geq$ 30, NM $\leq$ 3, Pmean $\geq$ 8, Pstd = 1, AltFwdReads $\geq$ 3, and AltRevReads $\geq$ 3. When calling low-frequency SNPs, it is difficult to remove all false positives without also removing some true positives. By treating all samples to the same sequence analysis pipeline, all samples will have a similar spectrum of false positives. By analyzing the ratios of different SNP types rather than raw SNP numbers, we further isolate biological effects from computational noise. VarDict was chosen because it is more sensitive to low allele-frequency variants [37].

DNA damage during library preparation was measured by individually analyzing the paired ends of Illumina paired-end sequencing and by looking for imbalances in mutations between the paired ends [47]. Mapped bam files were split into separate pairs, and GIV scores were calculated for each SNP type using the Damage-Estimator with mapping and base quality cutoffs set to 30.

Nuclear variants were called using Samtools mpileup (v. 1.3.1) and Bcftools call (v. 1.2) and were filtered for SNPgap of 3, Indelgap of 10, RPB > 0.1 and QUAL > 15, at least 3 high quality ALT reads (DP4(2) + DP4(3) ≥ 3), at least one high quality ALT read per strand (DP4(2) ≥ 1 and DP4(3) ≥ 1), and a high-quality ALT allele frequency ≥ 0.3. Chromosome 2 was excluded from this analysis to avoid false positives resulting from the presence of the large NuMT that has been duplicated and repeated there.

Supplementary Materials: The following are available online at http://www.mdpi.com/2223-7747/9/2/261/s1, Figure S1: Mitochondrial targeting of a GFP labeled MTP-APOBEC3G construct, Table S1: Primers for RT-PCR, Table S2: qPCR primer efficiency, Table S3: Primers for ROUS recombination assay, Supplementary File 2: SNP analysis tables.

Author Contributions: Conceptualization, E.W. and A.C.; methodology, E.W., E.P., and A.C.; software, E.W.; validation, E.W., E.P., and A.C.; formal analysis, E.W.; investigation, E.W., E.P. and A.C.; resources, A.C.; data curation, A.C.; writing—original draft preparation, E.W. and A.C.; writing—review and editing, E.W., E.P., and A.C.; visualization, E.W.; supervision, A.C.; project administration, A.C.; funding acquisition, A.C. All authors have read and agreed to the published version of the manuscript.

Funding: This research was funded by the National Science Foundation (USA), grants MCB-1413152 and MCB-1933590 to A.C.C.

Acknowledgments: Conversations with Arnie Bendich about organelle DNA replication and repair in meristem and vegetative cells were interesting and illuminating. We are grateful to Emily Jezewski for finding time in her busy golf schedule to do some of the qPCR experiments. Beth Rowan provided advice on Illumina library preparations from Arabidopsis leaves and many insights and helpful conversations about plant mitochondrial genomes. Daniel Sloan also provided insights and useful discussions. We thank Christian Elowski and the Nebraska Center for Biotechnology Core Research Facility for Microscopy for confocal fluorescent microscopy. This work used the Vincent J. Coates Genomics Sequencing Laboratory at UC Berkeley, supported by NIH S10 OD018174 Instrumentation Grant. Daniel Schachtman helped with disposal of leaf tissues from generations 2–9. The use of product and company names is necessary to accurately report the methods and results; however, the United States Department of Agriculture (USDA) neither guarantees nor warrants the standard of the products, and the use of names by the USDA implies no approval of the product to the exclusion of others that may also be suitable. The USDA is an equal opportunity provider and employer.

Conflicts of Interest: The authors declare no conflict of interest.

References

1. Drouin, G.; Daoud, H.; Xia, J. Relative rates of synonymous substitutions in the mitochondrial, chloroplast and nuclear genomes of seed plants. *Mol. Phylogenet. Evol.* **2008**, *49*, 827–831. [CrossRef]

2. Palmer, J.D.; Herbon, L.A. Plant mitochondrial DNA evolves rapidly in structure, but slowly in sequence. *J. Mol. Evol.* **1988**, *28*, 87–97. [CrossRef] [PubMed]

3. Richardson, A.O.; Rice, D.W.; Young, G.J.; Alverson, A.J.; Palmer, J.D. The "fossilized" mitochondrial genome of *Liriodendron tulipifera*: Ancestral gene content and order, ancestral editing sites, and extraordinarily low mutation rate. *BMC Biol.* **2013**, *11*, 29. [CrossRef] [PubMed]

4. Siekevitz, P. Powerhouse of the cell. *Sci. Am.* **1957**, *197*, 131–144. [CrossRef]

5. Wolfe, K.; Li, W.; Sharp, P. Rates of nucleotide substitution vary greatly among plant mitochondrial, chloroplast and nuclear DNAs. *Proc. Natl. Acad. Sci. USA* **1987**, *84*, 9054–9058. [CrossRef]

6. Sloan, D.B.; Alverson, A.J.; Chuckalovcak, J.P.; Wu, M.; McCauley, D.E.; Palmer, J.D.; Taylor, D.R. Rapid evolution of enormous, multichromosomal genomes in flowering plant mitochondria with exceptionally high mutation rates. *PLoS Biol.* **2012**, *10*, e1001241. [CrossRef]

7. Sloan, D.B.; Taylor, D.R. Testing for selection on synonymous sites in plant mitochondrial DNA: The role of codon bias and RNA editing. *J. Mol. Evol.* **2010**, *70*, 479–491. [CrossRef]

8. Wynn, E.L.; Christensen, A.C. Are synonymous substitutions in flowering plant mitochondria neutral? *J. Molec. Evol.* **2015**, *81*, 131–135. [CrossRef]

9. Kumar, R.A.; Oldenburg, D.J.; Bendich, A.J. Changes in DNA damage, molecular integrity, and copy number for plastid DNA and mitochondrial DNA during maize development. *J. Exp. Bot.* **2014**, *65*, 6425–6439. [CrossRef]

10. Segui-Simarro, J.M.; Coronado, M.J.; Staehelin, L.A. The mitochondrial cycle of Arabidopsis shoot apical meristem and leaf primordium meristematic cells is defined by a perinuclear tentaculate/cage-like mitochondrion. *Plant. Physiol.* **2008**, *148*, 1380–1393. [CrossRef]

11. Segui-Simarro, J.M.; Staehelin, L.A. Mitochondrial reticulation in shoot apical meristem cells of Arabidopsis provides a mechanism for homogenization of mtDNA prior to gamete formation. *Plant. Signal. Behav.* **2009**, *4*, 168–171. [CrossRef] [PubMed]

12. Boesch, P.; Ibrahim, N.; Paulus, F.; Cosset, A.; Tarasenko, V.; Dietrich, A. Plant mitochondria possess a short-patch base excision DNA repair pathway. *Nucleic Acids Res.* **2009**, *37*, 5690–5700. [CrossRef] [PubMed]

13. Gualberto, J.M.; Newton, K.J. Plant mitochondrial genomes: Dynamics and mechanisms of mutation. *Annu. Rev. Plant. Biol.* **2017**. [CrossRef] [PubMed]

14. Christensen, A.C. Genes and junk in plant mitochondria—Repair mechanisms and selection. *Genome Biol. Evol.* **2014**, *6*, 1448–1453. [CrossRef] [PubMed]

15. Christensen, A.C. Mitochondrial DNA repair and genome evolution. In *Annual Plant Reviews*, 2nd ed.; Logan, D.C., Ed.; Wiley-Blackwell: New York, NY, USA, 2018; Volume 50, Plant Mitochondria; pp. 11–31.

16. Cordoba-Cañero, D.; Dubois, E.; Ariza, R.R.; Doutriaux, M.P.; Roldan-Arjona, T. Arabidopsis uracil DNA glycosylase (UNG) is required for base excision repair of uracil and increases plant sensitivity to 5-fluorouracil. *J. Biol. Chem.* **2010**, *285*, 7475–7483. [CrossRef] [PubMed]

17. Krokan, H.E.; Standal, R.; Slupphaug, G. DNA glycosylases in the base excision repair of DNA. *Biochem. J.* **1997**, *325*, 1–16. [CrossRef]

18. Christensen, A.C. Plant mitochondrial genome evolution can be explained by DNA repair mechanisms. *Genome Biol. Evol.* **2013**, *5*, 1079–1086. [CrossRef]

19. Wynn, E.L.; Christensen, A.C. Repeats of unusual size in plant mitochondrial genomes: Identification, incidence and evolution. *G3 Genes Genomes Genet.* **2019**, *9*, 549–559. [CrossRef]

20. Klein, M.; Eckert-Ossenkopp, U.; Schmiedeberg, I.; Brandt, P.; Unseld, M.; Brennicke, A.; Schuster, W. Physical mapping of the mitochondrial genome of *Arabidopsis thaliana* by cosmid and YAC clones. *Plant. J.* **1994**, *6*, 447–455. [CrossRef]

21. Palmer, J.D.; Shields, C.R. Tripartite structure of the Brassica campestris mitochondrial genome. *Nature* **1984**, *307*, 437. [CrossRef]

22. Unseld, M.; Marienfeld, J.R.; Brandt, P.; Brennicke, A. The mitochondrial genome of *Arabidopsis thaliana* contains 57 genes in 366,924 nucleotides. *Nat. Genet.* **1997**, *15*, 57–61. [CrossRef] [PubMed]

23. Arrieta-Montiel, M.P.; Shedge, V.; Davila, J.; Christensen, A.C.; Mackenzie, S.A. Diversity of the arabidopsis mitochondrial genome occurs via nuclear-controlled recombination activity. *Genetics* **2009**, *183*, 1261–1268. [CrossRef] [PubMed]

24. Davila, J.I.; Arrieta-Montiel, M.P.; Wamboldt, Y.; Cao, J.; Hagmann, J.; Shedge, V.; Xu, Y.Z.; Weigel, D.; Mackenzie, S.A. Double-strand break repair processes drive evolution of the mitochondrial genome in Arabidopsis. *BMC Biol.* **2011**, *9*, 64. [CrossRef] [PubMed]

25. Abdelnoor, R.V.; Yule, R.; Elo, A.; Christensen, A.C.; Meyer-Gauen, G.; Mackenzie, S.A. Substoichiometric shifting in the plant mitochondrial genome is influenced by a gene homologous to MutS. *Proc. Natl. Acad. Sci. USA* **2003**, *100*, 5968–5973. [CrossRef]

26. Miller-Messmer, M.; Kuhn, K.; Bichara, M.; Le Ret, M.; Imbault, P.; Gualberto, J.M. RecA-dependent DNA repair results in increased heteroplasmy of the Arabidopsis mitochondrial genome. *Plant. Physiol.* **2012**, *159*, 211–226. [CrossRef]

27. Shedge, V.; Arrieta-Montiel, M.; Christensen, A.C.; Mackenzie, S.A. Plant mitochondrial recombination surveillance requires unusual RecA and MutS homologs. *Plant. Cell* **2007**, *19*, 1251–1264. [CrossRef]

28. Odahara, M.; Inouye, T.; Fujita, T.; Hasebe, M.; Sekine, Y. Involvement of mitochondrial-targeted RecA in the repair of mitochondrial DNA in the moss, Physcomitrella patens. *Genes Genet. Syst.* **2007**, *82*, 43–51. [CrossRef]

29. Abdelnoor, R.V.; Christensen, A.C.; Mohammed, S.; Munoz-Castillo, B.; Moriyama, H.; Mackenzie, S.A. Mitochondrial genome dynamics in plants and animals: Convergent gene fusions of a *MutS* homolog. *J. Molec. Evol.* **2006**, *63*, 165–173. [CrossRef]

30. Kleinstiver, B.P.; Wolfs, J.M.; Edgell, D.R. The monomeric GIY-YIG homing endonuclease I-BmoI uses a molecular anchor and a flexible tether to sequentially nick DNA. *Nucleic Acids Res.* **2013**, *41*, 5413–5427. [CrossRef]

31. Fukui, K.; Harada, A.; Wakamatsu, T.; Minobe, A.; Ohshita, K.; Ashiuchi, M.; Yano, T. The GIY-YIG endonuclease domain of Arabidopsis MutS homolog 1 specifically binds to branched DNA structures. *FEBS Lett.* **2018**, *592*, 4066–4077. [CrossRef]

32. Bendich, A.J. DNA abandonment and the mechanisms of uniparental inheritance of mitochondria and chloroplasts. *Chromosome Res. Int. J. Mol. Supramol. Evol. Asp. Chromosome Biol.* **2013**, *21*, 287–296. [CrossRef] [PubMed]

33. Chen, K.M.; Martemyanova, N.; Lu, Y.; Shindo, K.; Matsuo, H.; Harris, R.S. Extensive mutagenesis experiments corroborate a structural model for the DNA deaminase domain of APOBEC3G. *FEBS Lett.* **2007**, *581*, 4761–4766. [CrossRef] [PubMed]

34. Grefen, C.; Donald, N.; Hashimoto, K.; Kudla, J.; Schumacher, K.; Blatt, M.R. A ubiquitin-10 promoter-based vector set for fluorescent protein tagging facilitates temporal stability and native protein distribution in transient and stable expression studies. *Plant. J.* **2010**, *64*, 355–365. [CrossRef] [PubMed]

35. Li, H. Aligning sequence reads, clone sequences and assembly contigs with BWA-MEM. *arXiv* **2013**, arXiv:1303.3997.

36. Lai, Z.; Markovets, A.; Ahdesmaki, M.; Chapman, B.; Hofmann, O.; McEwen, R.; Johnson, J.; Dougherty, B.; Barrett, J.C.; Dry, J.R. VarDict: A novel and versatile variant caller for next-generation sequencing in cancer research. *Nucleic Acids Res.* **2016**, *44*, e108. [CrossRef]

37. Sandmann, S.; de Graaf, A.O.; Karimi, M.; van der Reijden, B.A.; Hellstrom-Lindberg, E.; Jansen, J.H.; Dugas, M. Evaluating variant calling tools for non-matched next-generation sequencing data. *Sci. Rep.* **2017**, *7*, 43169. [CrossRef]

38. Michalovova, M.; Vyskot, B.; Kejnovsky, E. Analysis of plastid and mitochondrial DNA insertions in the nucleus (NUPTs and NUMTs) of six plant species: Size, relative age and chromosomal localization. *Heredity* **2013**, *111*, 314–320. [CrossRef]

39. Richly, E.; Leister, D. NUMTs in sequenced eukaryotic genomes. *Mol. Biol. Evol.* **2004**, *21*, 1081–1084. [CrossRef]

40. Stupar, R.M.; Lilly, J.W.; Town, C.D.; Cheng, Z.; Kaul, S.; Buell, C.R.; Jiang, J. Complex mtDNA constitutes an approximate 620-kb insertion on Arabidopsis thaliana chromosome 2: Implication of potential sequencing errors caused by large-unit repeats. *Proc. Natl. Acad. Sci. USA* **2001**, *98*, 5099–5103. [CrossRef]

41. Halligan, D.L.; Keightley, P.D. Spontaneous mutation accumulation studies in evolutionary genetics. *Annu. Rev. Ecol. Evol. Syst.* **2009**, *40*, 151–172. [CrossRef]

42. Li, H. A statistical framework for SNP calling, mutation discovery, association mapping and population genetical parameter estimation from sequencing data. *Bioinformatics* **2011**, *27*, 2987–2993. [CrossRef] [PubMed]

43. Edmondson, A.C.; Song, D.; Alvarez, L.A.; Wall, M.K.; Almond, D.; McClellan, D.A.; Maxwell, A.; Nielsen, B.L. Characterization of a mitochondrially targeted single-stranded DNA-binding protein in Arabidopsis thaliana. *Mol. Genet. Genom.* **2005**, *273*, 115–122. [CrossRef] [PubMed]

44. Gualberto, J.M.; Mileshina, D.; Wallet, C.; Niazi, A.K.; Weber-Lotfi, F.; Dietrich, A. The plant mitochondrial genome: Dynamics and maintenance. *Biochimie* **2014**, *100*, 107–120. [CrossRef] [PubMed]

45. Khazi, F.R.; Edmondson, A.C.; Nielsen, B.L. An arabidopsis homologue of bacterial RecA that complements an E. coli recA deletion is targeted to plant mitochondria. *Mol. Genet. Genom.* **2003**, *269*, 454–463. [CrossRef] [PubMed]

46. Wallet, C.; Le Ret, M.; Bergdoll, M.; Bichara, M.; Dietrich, A.; Gualberto, J.M. The RECG1 DNA translocase is a key factor in recombination surveillance, repair, and segregation of the mitochondrial DNA in arabidopsis. *Plant. Cell* **2015**, *27*, 2907–2925. [CrossRef]

47. Chen, L.; Liu, P.; Evans, T.C.; Ettwiller, L.M. DNA damage is a pervasive cause of sequencing errors, directly confounding variant identification. *Science* **2017**, *355*, 752–756. [CrossRef]

48. Trasviña-Arenas, C.H.; Baruch-Torres, N.; Cordoba-Andrade, F.J.; Ayala-Garcia, V.M.; Garcia-Medel, P.L.; Diaz-Quezada, C.; Peralta-Castro, A.; Ordaz-Ortiz, J.J.; Brieba, L.G. Identification of a unique insertion in plant organellar DNA polymerases responsible for 5′-dRP lyase and strand-displacement activities: Implications for Base Excision Repair. *DNA Repair.* **2018**, *65*, 1–10. [CrossRef]

49. Schaack, S.; Ho, E.K.H.; Macrae, F. Disentangling the intertwined roles of mutation, selection and drift in the mitochondrial genome. *Philos. Trans. R. Soc. Lond. B Biol. Sci.* **2020**, *375*, 20190173. [CrossRef]

50. McGrew, D.; Knight, K. Molecular design and functional organization of the RecA protein. *Crit. Rev. Biochem. Mol. Biol.* **2003**, *38*, 385–432. [CrossRef]

51. Rowan, B.A.; Oldenburg, D.J.; Bendich, A.J. RecA maintains the integrity of chloroplast DNA molecules in Arabidopsis. *J. Exp. Bot.* **2010**, *61*, 2575–2588. [CrossRef]

52. Xu, L.; Marians, K. A dynamic RecA filament permits DNA polymerase-catalyzed extension of the invading strand in recombination intermediates. *J. Biol. Chem.* **2002**, *277*, 14321–14328. [CrossRef] [PubMed]

53. Eggler, A.L.; Lusetti, S.L.; Cox, M.M. The C terminus of the Escherichia coli RecA protein modulates the DNA binding competition with single-stranded DNA-binding protein. *J. Biol. Chem.* **2003**, *278*, 16389–16396. [CrossRef] [PubMed]

54. Garcia-Medel, P.L.; Baruch-Torres, N.; Peralta-Castro, A.; Trasvina-Arenas, C.H.; Torres-Larios, A.; Brieba, L.G. Plant organellar DNA polymerases repair double-stranded breaks by microhomology-mediated end-joining. *Nucleic Acids Res.* **2019**, *47*, 3028–3044. [CrossRef] [PubMed]

55. Zaegel, V.; Guermann, B.; Le Ret, M.; Andres, C.; Meyer, D.; Erhardt, M.; Canaday, J.; Gualberto, J.M.; Imbault, P. The plant-specific ssDNA binding protein OSB1 is involved in the stoichiometric transmission of mitochondrial DNA in Arabidopsis. *Plant. Cell* **2006**, *18*, 3548–3563. [CrossRef]

56. Backert, S.; Borner, T. Phage T4-like intermediates of DNA replication and recombination in the mitochondria of the higher plant *Chenopodium album* (L.). *Curr. Genet.* **2000**, *37*, 304–314. [CrossRef]

57. Rose, R.J.; McCurdy, D.W. New beginnings: Mitochondrial renewal by massive mitochondrial fusion. *Trends Plant. Sci.* **2017**, *22*, 641–643. [CrossRef]

58. Drake, J.W.; Baltz, R.H. The biochemistry of mutagenesis. *Annu. Rev. Biochem.* **1976**, *45*, 11–37. [CrossRef]

59. Lewis, C.A., Jr.; Crayle, J.; Zhou, S.; Swanstrom, R.; Wolfenden, R. Cytosine deamination and the precipitous decline of spontaneous mutation during Earth's history. *Proc. Natl. Acad. Sci. USA* **2016**, *113*, 8194–8199. [CrossRef]

60. Kauppila, J.H.K.; Bonekamp, N.A.; Mourier, A.; Isokallio, M.A.; Just, A.; Kauppila, T.E.S.; Stewart, J.B.; Larsson, N.G. Base-excision repair deficiency alone or combined with increased oxidative stress does not increase mtDNA point mutations in mice. *Nucleic Acids Res.* **2018**, *46*, 6642–6669. [CrossRef]

61. Carpenter, M.A.; Rajagurubandara, E.; Wijesinghe, P.; Bhagwat, A.S. Determinants of sequence-specificity within human AID and APOBEC3G. *DNA Repair.* **2010**, *9*, 579–587. [CrossRef]

62. Clough, S.J.; Bent, A.F. Floral dip: A simplified method for Agrobacteriummediated transformation of Arabidopsis thaliana. *Plant. J.* **1998**, *16*, 735–743. [CrossRef]

63. Karimi, M.; Inze, D.; Depicker, A. GATEWAY vectors for Agrobacterium-mediated plant transformation. *Trends Plant. Sci.* **2002**, *7*, 193–195. [CrossRef]

64. Onate-Sanchez, L.; Vicente-Carbajosa, J. DNA-free RNA isolation protocols for Arabidopsis thaliana, including seeds and siliques. *BMC Res. Notes* **2008**, *1*, 93. [CrossRef] [PubMed]

65. Bustin, S.A.; Benes, V.; Garson, J.A.; Hellemans, J.; Huggett, J.; Kubista, M.; Mueller, R.; Nolan, T.; Pfaffl, M.W.; Shipley, G.L.; et al. The MIQE guidelines: Minimum information for publication of quantitative real-time PCR experiments. *Clin. Chem.* **2009**, *55*, 611–622. [CrossRef] [PubMed]

66. Allen, G.C.; Flores-Vergara, M.A.; Krasynanski, S.; Kumar, S.; Thompson, W.F. A modified protocol for rapid DNA isolation from plant tissues using cetyltrimethylammonium bromide. *Nat. Protoc.* **2006**, *1*, 2320–2325. [CrossRef] [PubMed]

67. Rowan, B.A.; Patel, V.; Weigel, D.; Schneeberger, K. Rapid and inexpensive whole-genome genotyping-by-sequencing for crossover localization and fine-scale genetic mapping. *G3 Genes Genomes Genet.* **2015**, *5*, 385–398. [CrossRef] [PubMed]

68. Rowan, B.A.; Seymour, D.K.; Chae, E.; Lundberg, D.S.; Weigel, D. Methods for Genotyping-by-Sequencing. In *Genotyping: Methods and Protocols*; White, S.J., Cantsilieris, S., Eds.; Springer: New York, NY, USA, 2017; pp. 221–242. [CrossRef]

69. Caruccio, N. Preparation of next-generation sequencing libraries using Nextera technology: Simultaneous DNA fragmentation and adaptor tagging by in vitro transposition. *Methods Mol. Biol.* **2011**, *733*, 241–255. [CrossRef]

70. Baym, M.; Kryazhimskiy, S.; Lieberman, T.D.; Chung, H.; Desai, M.M.; Kishony, R. Inexpensive multiplexed library preparation for megabase-sized genomes. *PLoS ONE* **2015**, *10*, e0128036. [CrossRef]

71. Sloan, D.B.; Wu, Z.; Sharbrough, J. Correction of persistent errors in arabidopsis reference mitochondrial genomes. *Plant. Cell* **2018**, *30*, 525–527. [CrossRef]

72. Berardini, T.Z.; Reiser, L.; Li, D.; Mezheritsky, Y.; Muller, R.; Strait, E.; Huala, E. The arabidopsis information resource: Making and mining the "gold standard" annotated reference plant genome. *Genesis* **2015**, *53*, 474–485. [CrossRef]

73. Li, H.; Handsaker, B.; Wysoker, A.; Fennell, T.; Ruan, J.; Homer, N.; Marth, G.; Abecasis, G.; Durbin, R. The sequence alignment/map format and samtools. *Bioinformatics* **2009**, *25*, 2078–2079. [CrossRef] [PubMed]

74. McKenna, A.; Hanna, M.; Banks, E.; Sivachenko, A.; Cibulskis, K.; Kernytsky, A.; Garimella, K.; Altshuler, D.; Gabriel, S.; Daly, M.; et al. The genome analysis toolkit: A MapReduce framework for analyzing next-generation DNA sequencing data. *Genome Res.* **2010**, *20*, 1297–1303. [CrossRef] [PubMed]

Article

De Novo Assembly Discovered Novel Structures in Genome of Plastids and Revealed Divergent Inverted Repeats in *Mammillaria* (Cactaceae, Caryophyllales)

Sofía Solórzano [1,*], Delil A. Chincoya [1], Alejandro Sanchez-Flores [2,*], Karel Estrada [2], Clara E. Díaz-Velásquez [3], Antonio González-Rodríguez [4], Felipe Vaca-Paniagua [3,5], Patricia Dávila [6] and Salvador Arias [7]

[1] Laboratorio de Ecología Molecular y Evolución, UBIPRO, FES Iztacala, Universidad Nacional Autónoma de México, Avenida de los Barrios 1, Los Reyes Iztacala, Tlalnepantla de Baz 54090, Estado de México, Mexico; dela@comunidad.unam.mx

[2] Unidad Universitaria de Secuenciación Masiva y Bioinformática, Instituto de Biotecnología, Universidad Nacional Autónoma de México, Avenida Universidad 2001, Chamilpa, Cuernavaca 62250, Mexico; karel@ibt.unam.mx

[3] Laboratorio Nacional en Salud: Diagnóstico Molecular y Efecto Ambiental en Enfermedades Crónico-Degenerativas, FES Iztacala, Universidad Nacional Autónoma de México, Los Reyes Iztacala, Tlalnepantla de Baz 54090, Estado de México, Mexico; cdiaz@comunidad.unam.mx (C.E.D.-V.); Felipe.vaca@iztacala.unam.mx (F.V.-P.)

[4] Laboratorio de Genética de la Conservación, Instituto de Investigaciones en Ecosistemas y Sustentabilidad, Universidad Nacional Autónoma de México, Antigua carretera a Pátzcuaro 8701, Ex-Hacienda San José La Huerta, Morelia 58190, Michoacán, Mexico; agrodrig@cieco.unam.mx

[5] Subdirección de Investigación Básica, Instituto Nacional de Cancerología, Ciudad de México 04510, Mexico

[6] Laboratorio de Recursos Naturales, UBIPRO, FES Iztacala, Universidad Nacional Autónoma de México, Avenida de los Barrios 1, Los Reyes Iztacala, Tlalnepantla de Baz 54090, Estado de México, Mexico; pdavilaa@unam.mx

[7] Jardín Botánico, Instituto de Biología, Universidad Nacional Autónoma de México, Tercer Circuito Exterior, Ciudad Universitaria, Coyoacán, Ciudad de México 04510, Mexico; sarias@ib.unam.mx

* Correspondence: solorzanols@unam.mx (S.S.); alexsf@ibt.unam.mx (A.S.-F.)

Received: 21 August 2019; Accepted: 22 September 2019; Published: 1 October 2019

Abstract: The complete sequence of chloroplast genome (cpDNA) has been documented for single large columnar species of Cactaceae, lacking inverted repeats (IRs). We sequenced cpDNA for seven species of the short-globose cacti of *Mammillaria* and de novo assembly revealed three novel structures in land plants. These structures have a large single copy (LSC) that is 2.5 to 10 times larger than the small single copy (SSC), and two IRs that contain strong differences in length and gene composition. Structure 1 is distinguished by short IRs of <1 kb composed by *rpl23-trnI-CAU-ycf2*; with a total length of 110,189 bp and 113 genes. In structure 2, each IR is approximately 7.2 kb and is composed of 11 genes and one Intergenic Spacer-(*psbK-trnQ*)-*trnQ-UUG-rps16-trnK-UUU-matK-trnK-UUU-psbA-trnH-GUG-rpl2-rpl23-trnI-CAU-ycf2*; with a total size of 116,175 bp and 120 genes. Structure 3 has divergent IRs of approximately 14.1 kb, where IRA is composed of 20 genes: *psbA-trnH-GUG-rpl23-trnI-CAU-ycf2-ndhB-rps7-rps12-trnV-GAC-rrn16-ycf68-trnI-GAU-trnA-AGC-rrn23 -rrn4.5-rrn5-trnR-ACG-trnN-GUU-ndhF-rpl32*; and IRB is identical to the IRA, but lacks *rpl23*. This structure has 131 genes and, by pseudogenization, it is shown to have the shortest cpDNA, of just 107,343 bp. Our findings show that *Mammillaria* bears an unusual structural diversity of cpDNA, which supports the elucidation of the evolutionary processes involved in cacti lineages.

Keywords: divergent inverted repeats; short-globose cacti; novel gene rearrangements; pseudogenization

---

1. Introduction

A new era in the study of evolutionary processes of chloroplasts and their genomes has arisen with the advent of massive sequencing [1]. Huge advances have been documented since 1883, when Schimper postulated an endosymbiotic cyanobacterial origin of these organelles [2]. More recently, many studies have focused on determining the cyanobacterial origin of the DNA molecule contained in chloroplasts [1,3,4]. Using comparative genomics, DNA sequences of complete genomes of contemporary cyanobacteria, algae, and plants have been analyzed, leading to the discovery that the chloroplast genome encompasses structural changes with significant evolutionary information. Thus, in comparison to cyanobacteria and algae, a significant reduction in the total length and in the number of genes has been documented in land plants [5]. However, many genes lacking in the chloroplast genome have migrated to nuclear or mitochondrial genomes [6], which indicates a complex functional relationship among the three genomes contained in plants. In addition, in plants, the chloroplast genome has a hybrid transcriptional process, which denotes the evolutionary transition from a prokaryotic form to a eukaryotic form. Accordingly, most encoding regions are regulated in operons which are transcribed into polycistronic units, as occurs in contemporary cyanobacteria [7]. Additionally, typical eukaryotic transcriptional regulation was documented in nearly 60 promoters for encoding regions and their transfer RNAs [8].

Comparisons within land plants have concluded that the differences in the total length of the complete chloroplast genome (cpDNA) are caused by lengthening or shortening of genes and not by a significant gain/loss of them [5,9]. Flowering land plants tend to have a total of 120 genes; of these, nearly 80 are encoding genes, 30 are tRNAs, and four are rRNAs [9]. In angiosperms, these genes are not randomly distributed along the entire molecule of DNA of the chloroplasts, but instead determine a recognizable structure in the cpDNA. Moreover, in most angiosperms, this cpDNA is sectioned into four regions, which are distinguished by their length and trend in gene composition. The largest single copy (LSC) contains most of the encoding genes directly related to photosystems I and II, ATP synthases, proteins of cytochrome b/f complex, DNA depending on RNA polymerases, and proteins which tend to have a single encoding gene: ribulose bisphosphate carboxylase large chain, maturase K, envelope membrane protein, acetyl coenzyme carboxylase, transcriptional initiation factor, as well as most of proteins of small and large subunits of the ribosome. The small single copy (SSC) often contains dehydrogenase subunits and open reading frames. This copy typically shows the highest mutation rates. The SSC is often flanked by two inverted repeats (IRA and IRB), which vary in gene composition and length [1]. In these plants, the IRs typically contains four ribosomal RNA subunits (*4.5S, 5S, 16S,* and *23S*) and five transfer RNA subunits (*trnA-UGC, trnI-GAU, trnN-GUU, trnR-ACG,* and *trnV-GAC*). In addition, the IRs exhibit lower mutation rates than the SSC [10]. Currently, nearly 500 cpDNA have been sequenced for the land plant group, showing that IRs are the main source of structural variation by relative expansion, contraction, and gene rearrangement. However, between IRA and IRB within the same genome, there are no differences in gene arrangement and composition, and in only a few cases do they have low divergence in the DNA sequence [9,10].

In angiosperms, although IRs are commonly present, they are absent in some taxa. Around 95% of legume species of the subfamily Papilionoideae (order Fabales) lack IRs, which has been interpreted as a novel evolutionary change that appeared in a common ancestor and, eventually, was inherited by its descendants, whereas other legume species of this order have IRs [11]. Recently, the lack of IRs was documented in two species of large columnar cacti of Cactoideae (Cactaceae, Caryophyllales) of the tribe Echinocereeae. The loss of IRs in the saguaro (*Carnegiea gigantea*) was interpreted as a novel structural change [12]. In addition, we have verified that the cpDNA of *Pachycereus schottii*, which has recently been directly submitted to the GenBank database, also lacks IRs (uploaded with its synonym *Lophocereus schottii*, NCBI, NC_041727.1). In contrast, in all other species currently sequenced in Caryophyllales have been shown to have IRs [13]. At the infrageneric level, contrasting results have been documented and it is not a rule that all members of a certain genus show identical cpDNA structure. For example, in 13 species of *Camellia* (Theaceae, Ericales), identical structure of cpDNA and low divergence of DNA sequences were documented [14]. A similar result was obtained for seven species of *Silene* (Caryophyllaceae, Caryophyllales), with identical cpDNA structure and only a small gain/loss of genes among them being documented [15,16]. In contrast, unusual results have been obtained for 17 species of *Erodium* (Geraniaceae, Geraniales), which showed deep and strong structural changes, such as expansion and contraction of IRs or even the absence of IRs, and substantial gene rearrangements in the LSC [17]. Thus, data based on characterizations of the structures of complete chloroplast genomes are necessary, as they might reveal novel unexpected results that may help to clarify evolutionary processes in plants.

In this study, we focused on cacti species of the short-globose genus *Mammillaria* (Cactoideae, tribe Cacteae). *Mammillaria* is relevant, in terms of biodiversity, due to its high species richness (163–232) [18] in the Cactaceae. A total of 192 species and subspecies of *Mammillaria* are listed in the Red List of Threatened Species of International Union for Conservation of Nature [19]. For the species of this genus, non-fully resolved phylogenies were obtained from DNA sequences of the *rpl16* intron and psbA-trnH intergenic spacer regions of the chloroplast [20]. The increment of plastid molecular markers (*rpl16*, *trnK*, and *rpoC1* introns, and *trnK-psbA*, *rpl20-rps12*, *trnL-trnF*, and *trnT-trnL* intergenic spacers) did not resolve the relationships among species of *Mammillaria*, nor of species from closer genera (i.e., *Coryphantha*, *Escobaria*, *Neolloydia*, *Ortegocactus*, and *Pelecyphora*). These unresolved evolutionary relationships have been attributed to the recent origin of Cactaceae (e.g., [21,22]), estimated at 35 million years ago [23]. Currently, morphological characteristics have been used to postulate the taxonomic limits among species of *Mammillaria* and of those in close cacti genera [18]. However, these characters are ambiguous and often do not accomplish a robust taxonomically resolved separation [20,21].

In this study, we de novo assembled the complete chloroplast genome of seven species in this genus, in order to utilize these genomes as reference for *Mammillaria*. A second objective was to identify putative structural characteristics of the cpDNA of *Mammillaria*, by comparing with the complete chloroplast genomes, which have been documented for other cacti species and other Caryophyllales. In addition, we discuss whether the structural differences of the cpDNA discovered in *Mammillaria* may serve to resolve the evolutionary and taxonomic pendants of this genus. As structural differences have been documented at the subfamily level, we expect the cpDNA of *Mammillaria* (Cacteae) to differ from those of the large columnar cacti (Echinocereeae); however, among species of *Mammillaria*, structural differences in cpDNA are not expected by its recent divergence.

2. Results

2.1. Gene Composition and Length Variation in Three Novel cpDNA Structures Identified in Mammillaria

De novo assembly revealed three structures of cpDNA in *Mammillaria* (Figure 1). Structure 1 was present in *M. albiflora* and *M. pectinifera* (Figure 1a), structure 2 in *M. crucigera*, *M. huitzilopochtli*, *M. solisioides*, and *M. supertexta* (Figure 1b), and structure 3 in *M. zephyranthoides* (Figure 1c). These structures had a quadripartite partition, into LSC, SSC, and two IRs (Figure 1). We identified unexpected and strong structural differences in gene composition and length among the three structures (Figures 2 and 3a). In addition, structure 3 (*M. zephyranthoides*) had divergent IRs; meanwhile, the rest of the species had identical gene composition in their IRs (Figure 1).

Variation between species in the total length and number of genes of cpDNA were documented (Tables 1 and 2). The cpDNA of *Mammillaria* ranged from 107,343 bp (*M. zephyranthoides*) to 116,175 bp (*M. supertexta*) (Table 1). The relative length of LSC represented approximately 62% of the genome in the six species (*M. albiflora*, *M. crucigera*, *M. huitzilopochtli*, *M. pectinifera*, *M. solisioides*, and *M. supertexta*) (Table 1). Moreover, the LSC was nearly 2.5 times larger than SSC, which represented 25.35–28% of the total genome length (Table 1). In contrast, in *M. zephyranthoides*, the LSC was longer (68% of the cpDNA length), being 10 times larger than its respective SSC, whose length only reached 7%. The shortening of SSC in *M. zephyranthoides* was due to the lengthening of the IRs, which had nearly 14 kb, and to the strong reduction of the genes *ycf1* and *ycf2* to <1 kb; meanwhile, in the other species of *Mammillaria*, these genes were >6 kb (Figure 1, Table 1).

We identified additional different types of gene rearrangements at the LSC (Figure 3b). The first was a type of rearrangement involving blocks of genes, which were inverted but maintained identical order (Figure 3b); there was also a second type involving a single gene with two variants: (a) the single gene did not change its relative location, but its orientation was inverted (*trnF-GAA*); (b) the single gene changed in location, but it maintained its orientation (*rpl2*). The single gene *trnF-GAA* had identical orientation in the species of structure 1 (*M. albiflora* and *M. pectinifera*), structure 3 (*M. zephyranthoides*), and *M. solisioides* of structure 2, but was inverted in the other three species of structure 2 (*M. crucigera*, *M. huitzilopochtli*, and *M. supertexta*). In the structure 1 species, *rpl2* flanked the IRA in *M. pectinifera*, whereas, in *M. albiflora*, it flanked the IRB. In addition, in *M. supertexta* (structure 2) and *M. zephyranthoides* (structure 3), *rpl33* was lost, but it was present in the other five species of *Mammillaria* as pseudogene except in *M. albiflora* (Figure 1, Table 2).

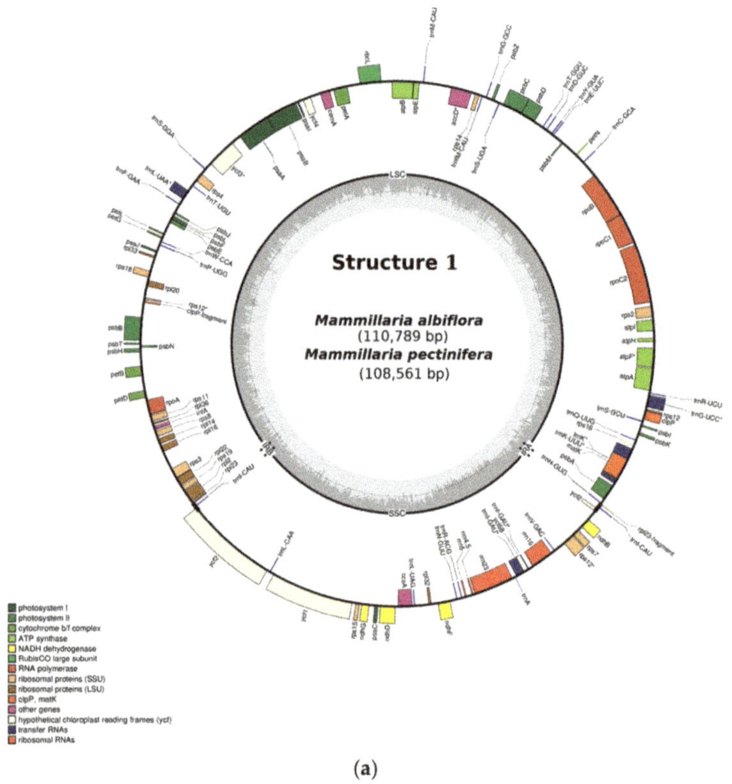

(a)

Figure 1. *Cont.*

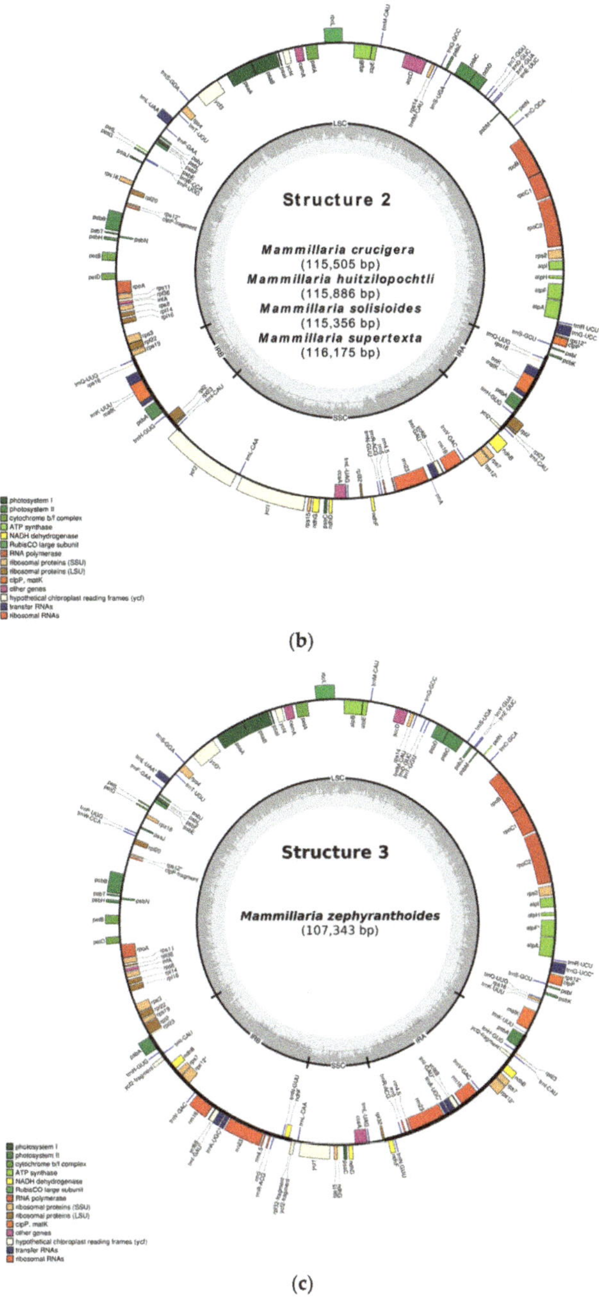

(b)

(c)

Figure 1. Three different structures found in the complete chloroplast genome of *Mammillaria*: (**a**) structure 1, (**b**) structure 2, and (**c**) structure 3. In structure 1, the *rpl2* gene is flanking IRB in *M. albiflora* and IRA in *M. pectinifera*. Gene *rpl33* was lost in *M. supertexta* of structure 2 and in *M. zephyranthoides* of structure 3. The genomes are displayed circularly, and IRA and IRB correspond to duplicated blocks of regions; starting from the top of the circle, the IRA is the one that appears first in clockwise.

On the other hand, comparison of the complete genomes of the seven species showed similar percentages of types of genes. In the three structures, the highest percentage of genes corresponded to tRNAs (26%), where each of the sets of *rps* and *psb* represented 13% of the genes. Another similarity found among species was that, in the LSC, large blocks of concatenated genes maintained identical gene compositions and arrangements (Figures 1 and 2). Most of the concatenated genes correspond to the encoding genes of photosystems I and II. In addition, the *rbcL* gene, units of the cytochrome b/f complex, and genes of the DNA-dependent RNA polymerase (Table 2) were identical in number, location, and arrangement (Figure 1). Comparisons of LSC and SSC showed that structures 1 and 2 had more mutual similarities than either had with structure 3 (Table 2). In the SSC, the seven species maintained identical order and orientation in *ycf2-trnL-CAA-ycf1*; although, in structure 3, they were shorter by pseudogenization (Figure 1).

In addition, in the seven species pseudogenization was identified in the NADH dehydrogenase-like (NDH) complex of plastid genes (*ndh* genes, hereafter). Of this family of genes, only four subunits of the suite of dehydrogenase genes (B, D, F, and G) were recorded in the seven species. The *ndh* subunits B, D, and F were pseudogenes in the seven species, and subunit G was pseudogene only in *M. pectinifera*, *M. solisioides* and *M. zephyranthoides*. In addition, *ycf68* was a pseudogene in all seven species, and *ycf4* was pseudogene in three species except in *M. albiflora*, *M. huitzilopochtli*, *M. supertexta* and *M. zephyranthoides* (Table 2).

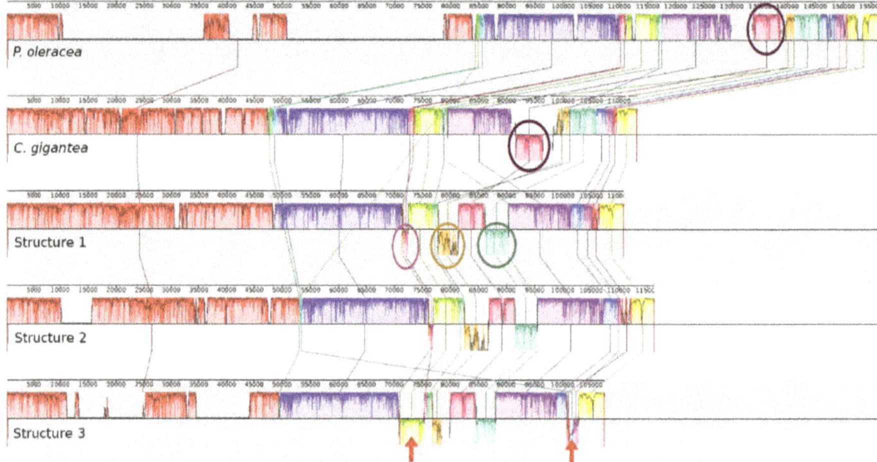

Figure 2. MAUVE graphic of five structural alignments of complete chloroplast genomes. The upper graph corresponds to caryophyllid *P. oleracea* (Portulacaceae); below that, the large giant columnar cactus, *C. gigantea*; and the last three graphs are the three structures documented in *Mammillaria*. Relative inverted DNA sequences are drawn above/below of the horizontal line; identical genes are in the same color. *P. oleracea* has a larger genome than any species of Cactaceae. Discarding the IRs that are recorded in *Mammillaria* and *P. oleracea*, but not in *C. gigantea*, the cpDNA structure of *P. oleracea* is more similar in structure to *C. gigantea* than to *Mammillaria*. Between *C. gigantea* and *P. oleracea*, a single large block of inverted genes (encircled) corresponding to *atpB* and *atpE* is shown. This block of genes in *Mammillaria* has identical orientation to *P. oleracea*. In *Mammillaria*, many other novel gene rearrangements, which are absent in the other two-caryophyllid taxa, were documented. Additionally, structure 3 has two blocks of inverted genes (described in detail in Figure 3b), with respect to structures 1 and 2. These two blocks of genes are indicated with arrows and have identical orientation in *C. gigantea*, *P. oleracea*, and structures 1 and 2 of *Mammillaria*.

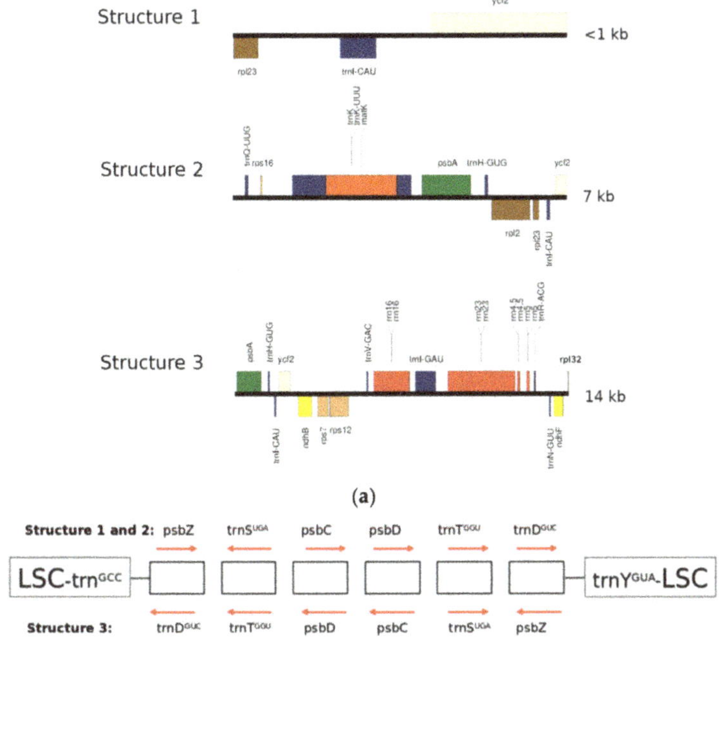

(a)

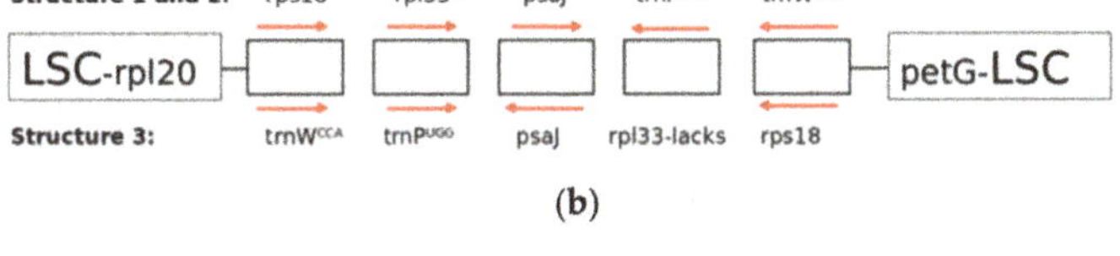

(b)

Figure 3. (a) Comparison of length and gene composition of IRs in the three structures documented for the complete chloroplast genomes of *Mammillaria*. The two IRs of structure 3 diverge in *rpl23*; its location in IRA is denoted with an asterisk. (b) Blocks of genes rearranged at the LSC. These genes are inverted and reoriented in structures 1 and 2, with respect to structure 3. The direction of the row indicates the orientation of transcription, to the left in sense of clockwise and to the right, counter-clockwise. The large squares indicate the genes of LSC that flank these two rearrangements. The asterisk in *rpl33* (bottom figure) indicates that, in *M. supertexta* of structure 2 and in species of structure 3, this gene was lost.

Table 1. Species of *Mammillaria* grouped by the type of the structure identified in the complete chloroplast genome (cpDNA). Within and among structure variation in total length size, the two inverted repeats (IRs), large single copy (LSC), and small single copy (SSC) were detected.

Type of Structure	Total Length	IRs	LSC	SCC	Total Number of Genes	Access Number [1]
I. Structure 1						
1.1 *M. albiflora*	110789	1348	78380	31061	113	MN517610
1.2. *M. pectinifera*	108561	1544	72273	29744	113	MN519716
II. Structure 2						
1. *M. crucigera*	115505	14522	71565	29418	120	MN517613
2. *M. huitzilopochtli*	115886	14488	71997	29401	120	MN517612
3. *M. solisioides*	115356	14428	71690	29238	120	MN518341
4. *M. supertexta*	116175	14490	72240	29445	119	MN508963
III. Structure 3						
1. *M. zephyranthoides*	107343	28252	71811	7281	131	MN517611

[1] GeneBank access number of the DNA sequences deposited.

Table 2. Variation in structural and functional gene composition in the three structures of cpDNA found in *Mammillaria*. A total of 18 different types of genes were documented, and these are organized alphabetically according to their location in IRs, LSC, and SSC. All the genes located at IRs are duplicated (2X), except the *rpl23*Ψ in structure 3 that lacks in IRB.

Gene Type/Structure	Region	Structure 1	Structure 2	Structure 3
1. Ribosomal RNA (rrn)	SSC	rrn4.5, 5,16, 23	rrn4.5, 5,16, 23	
	IRs			rrn4.5, 5,16, 23 (2X)
2. Transfer RNA (trn)	LSC	$trnC^{GCA}$, $trnD^{GUC}$, $trnE^{UUC}$, $trnF^{GAA}$, $trnG^{GCC}$, $trnG^{UCC}$, $trnH^{GUG}$, $trnK^{UUU}$, $trnL^{UAA}$, $trnM^{CAU}$, $trnM^{CAU}$, $trnP^{UGG}$, $trnQ^{UUG}$, $trnR^{UCU}$, $trnS^{GGA}$, $trnS^{GGU}$, $trnS^{UGA}$, $trnT^{GGU}$, $trnT^{UGU}$, $trnY^{GUA}$	$trnC^{GCA}$, $trnD^{GUC}$, $trnE^{UUC}$, $trnF^{GAA}$, $trnG^{GCC}$, $trnG^{UCC}$, $trnL^{UAA}$, $trnM^{CAU}$, $trnfM^{CAU}$, $trnP^{UGG}$, $trnR^{UCU}$, $trnS^{GGA}$, $trnS^{GCU}$, $trnS^{UGA}$, $trnT^{GGU}$, $trnT^{UGU}$, $trnW^{CCA}$, $trnY^{GUA}$	$trnC^{GCA}$, $trnD^{GUC}$, $trnE^{UUC}$, $trnF^{GAA}$, $trnG^{GCC}$, $trnG^{UCC}$, $trnK^{UUU}$, $trnL^{UAA}$, $trnM^{CAU}$, $trnfM^{CAU}$, $trnP^{UGG}$, $trnQ^{UUG}$, $trnR^{UCU}$, $trnS^{GGA}$, $trnS^{GCU}$, $trnS^{UGA}$, $trnT^{GGU}$, $trnT^{UGU}$, $trnW^{CCA}$, $trnY^{GUA}$
	SSC	trnA-f, $trnI^{GAU}$, $trnI^{GAU}$, $trnL^{CAA}$, $trnL^{UAG}$, $trnN^{GUU}$, $trnR^{ACG}$, $trnV^{GAG}$	trnA-f, $trnI^{GAU}$, $trnI^{UAG}$, $trnN^{GUU}$, $trnL^{CAA}$, $trnR^{ACG}$, $trnV^{GAG}$	$trnL^{UAG}$, $trnL^{CAA}$
	IRs	$trnI^{CAU}$ (2X)	$trnH^{GUG}$, tmI^{CAU}, $trnK^{UUU}$, $trnQ^{UUG}$ (2X)	$trnA^{UGC}$, $trnH^{GUG}$, $trnI^{CAU}$, $trnI^{GAU}$, $trnN^{GUU}$, $trnR^{ACG}$, $trnV^{GAC}$ (2X)
3. Proteins of small subunits of the ribosome (rps)	LSC	rps2, 3, 4, 8, 11, 12 (2), 14, 16Ψ, 18Ψ, 19	rps2, 3, 4, 8, 11, 12 (2), 14, 18Ψ, 19	rps2, 3, 4, 8 11, 12, 12Ψ, 14, 16Ψ, 18Ψ, 19
	SSC	rps7, 12, 15	rps7, 12, 15	rps15
	IRs		rps16Ψ (2X)	rps7, 12, (2X)

Table 2. Cont.

Gene Type/Structure	Region	Structure 1	Structure 2	Structure 3
4. Proteins of large subunits of the ribosome (rpl)	LSC	rpl2, 14, 16, 20, 22, 33Ψ, 36Ψ	rpl14, 16Ψ, 20, 22,33Ψ*, 36Ψ	rpl2, 14, 16Ψ, 20, 22, 23Ψ, 36
	SSC	rpl32	rpl32	
	IRs	rpl23Ψ (2X)	rpl2, 23Ψ (2X)	rpl32 (2X), 23Ψ (IRA)
5. DNA dependent RNA polymerase (rpo)	LSC	rpoA, B, C1, C2	rpoA, B, C1, C2	rpoA, B, C1, C2,
6. NADH dehydrogenase (ndh)	SSC	ndhBΨ, DΨ, FΨ, GΨ ***	ndhBΨ, DΨ, FΨ, G ***	
	IRs			ndhBΨ, DΨ, FΨ, GΨ (2X)
7. Photosystem I (psa)	LSC	psaA, B, I, J	psaA, B, I, J	psaA, B, I, J
	SSC	psaC	psaC	psaC
8. Photosystem II (psb)	LSC	psbA, B, C, D, E, F, H, I, J, K, L, M, N, T, Z	psbB, C, D, E, F, H, I, J, K, L, M, N, T, Z	psbB, C, D, E, F, H, I, J, L, K, M, N T, Z
	IRs		psbA (2X)	psbA (2X)
9. Cytochrome b/f complex (pet)	LSC	petA, B, D, G, L, N	petA, B, D, G, L, N	petA, B, D, G, L, N
10. ATP synthase (atp)	LSC	atpA, B, E, F, H, I	atpA, B, E, F, H, I	atpA, B, E, F, H, I
11. Rubisco (rbc)	LSC	rbcL	rbcL	rbcL
12. Maturase K	LSC	matK		matK
	IRs		matK (2X)	
13. Protease (clp)	LSC	clpPΨ, clpP	clpPΨ, clpP	clpPΨ, clpP
14. Envelope membrane protein (cem)	LSC	cemA	cemA	cemA
15. Subunit of acetil-CoA-carboxylase (acc)	LSC	accDΨ	accDΨ	accDΨ
16. c-type cytochrome synthesis (ccs)	SSC	ccsA	**SSC**: ccsA	**SSC**: ccsA
17. Translational initiation factor (inf)	LSC	infA	infA	infA
18. Hypothetical chloroplast reading frames (ycf)	LSC	ycf3, ycf4Ψ	ycf3, ycf4Ψ **	ycf3, ycf4
	SSC	ycf1, ycf2, ycf68Ψ	ycf1, ycf2, ycf68Ψ	ycf1Ψ, ycf2Ψ
	IRs	ycf2-p (2X)	ycf2-p (2X)	ycf2Ψ, ycf68Ψ (2X)

Ψ indicates a pseudogene. The note "-p" indicates that a partial DNA sequence of a gene is inserted in the two IRs. * indicates that *rpl33* lacks in *M. supertexta* but it is present in the other three species of this structure. In addition, this gen is pseudogene in all species except in *M. albiflora*. ** indicates that is a pseudogene in *M. crucigera* and *M. solisioides* of structure 2. *** indicates that it is pseudogene only in *M. solisioides* of structure 2 and *M. pectinifera* of structure 1.

2.2. Structure 1: Shortest IRs, Composed of Three Genes, rpl23-trnI-CAU-ycf2

This structure was distinguished by two short IRs (of <1 kb) composed by *rpl23-trnI-CAU-ycf2*. Of this *ycf2*, a total of 265 bases of its 5′ extreme are inserted in the IRs, which were identical in DNA sequence and gene composition. This structure 1 was found only in two species of *Krainzia* subgenus, *M. albiflora*, and *M. pectinifera* (Figure 1, Table 1). The genome of *M. albiflora* was larger (110,789 bp, Table 1) than the one of *M. pectinifera*. In this structure, both IRs had length 1348 bp (1.22% of the total genome). The three genes represented 5.3% of the total genes (113). These two species had an identical number of total genes; most of which (26.6%) were represented by tRNAs (Figure 1a, Table 1). The LSC covered the largest proportion of the DNA sequence (72.6%, Table 1) and the largest number of genes (82) (Table 2). However, *rpl2* gene is flaking IRA in *M. pectinifera* but in *M. albiflora* is flanking IRB that is the identical location of the species of structures 2 and 3.

2.3. Structure 2: IRs Composed by an Unusual Complete Battery of 11 Concatenated Genes and One Identical Intergenic Spacer.

Structure 2 was found in three species of the subgenus *Mammillaria* (M. crucigera, M. Huitzilopochtli, and *M. supertexta*) and one of the subgenus *Krainzia* (M. solisioides). In all of these, the two IRs were formed by genes usually located at the LSC, and were flanked by the DNA sequence of an identical intergenic spacer sequence (IGS) that is from psbk to trnQ; however, IRB was flanked by *rps19*, whereas IRA was flanked by *psbK* (Figure 1). The complete composition of this structure consisted of this IGS and 11 genes: IGS (*psbK-trnQ*)-*trnQ-UUG-rps16-trnK-UUU-matK-trnK -UUU-psbA-trnH-GUG-rpl2-rpl23-trnI-CAU-ycf2*. The four species with structure 2 had identical IRs, with respect to gene composition and DNA sequence. However, there were differences between species, in terms of the total genome length, number of genes, and length of each of the four quadripartite regions (Table 1). The *rpl33* gene was found in three species, but was absent in *M. supertexta*. Consequently, the total number of genes differed among the structure 2 species (Tables 1 and 2). Although they showed differences, the four species had similar percentages in the relative proportions of genes represented in IRs, LSC, and SSC, as well as in the percentages of gene types in the overall genome (Tables 1 and 2). Using *M. supertexta*, as a reference is important, as it had the largest genome in structure 2, showing 63% (75) of genes located at the LSC and 20.2% (24) at the SSC. In addition, each IR comprised 6.24% of the genes (11 genes and one IGS). In structure 2, the complete gene of maturase K gen (*matK*) is nested with *trnK-UUU* introns and inserted into the IRs. In this structure, 28.20% were tRNAs, followed by 13.67% in each of the suites of *rps* and *psb* subunits, *rpl* subunits represent 8.40%. Finally, the DNA regions that were represented by only one gene (0.85%) were *accD*, *ccsA*, *cemA*, *infA*, and *rbcL* (Table 2).

2.4. Structure 3: Largest and Divergent IRs in Which Four Ribosomal Units are Included.

Structure 3 was only recorded in *M. zephyranthoides* (Figure 1c). This genome had a length of 107,343 bp and 131 genes (Table 1). The LSC covered 62% of the total number of genes (81), the SSC had only 7 genes (5.7%), and both IRs had 43 genes (32.8%). In addition, each IR was approximately 14.1 kb in length and was comprised of genes typically located at the SSC. The IRA was composed of *psbA-trnH-GUG-rpl23*-trnI-CAU-ycf2partial-ndhB-rps7-rps12-trnV-GAC-rrn16-ycf68-trnI-GAU-trnA-AGC -rrn23-rrn4.5-rrn5-trnR-ACG-trnN-GUU-ndhF-rpl32*. The *rpl23* is marked with an asterisk because it is absent in the IRB, thus the IRB was identical in gene composition and arrangements but was divergent. Structure 3 is distinguished by the ancestral presence of four rRNA subunits (4.5, 5, 16, and 23) in the IRs.

Pseudogenization in structure 3 was clearly identified, in that *ycf2* and *ycf1* showed incomplete DNA sequences. The former was truncated into three segments. Two of these segments were inserted into the IRS and the third one was at the SSC. These three segments added only to a total of 960 bp. The gene *ycf1* had <1000 bp. Consequently, the shortening of *ycf1* and *ycf2* caused a diminished cpDNA total length (Table 1, Figure 1c). Additionally, this structure added, as pseudogenes, *accd*, *rps18*, *rpl23*, and one copy of *rps12*.

3. Discussion

The three structures of cpDNA discovered in *Mammillaria* are novel and these have not been recorded in other eukaryote organisms. In addition, the divergent IRs (identified in structure 3) are a novel result for land plants (Embryophyta), only having otherwise been recently discovered in green algae of the order Ignatiales (*Pseudoneochloris*, and *Chamaetrichon*) [24]. These strong arrangements in the cpDNA of *Mammillaria* are notable with respect to the rest of caryophyllids as *P. oleracea* (Portulacaceae) and *C. gigantea* (Cactaceae), and even within *Mammillaria* (Figures 2 and 3). Based on the DNA sequences of chloroplasts and current biogeographic distribution, it was estimated that the suborder Portulacinae, which includes the families Cactaceae and Portulacaceae, diverged in the early Miocene (18.8–33.7 Mya). In particular, for Cactaceae, an origin of 10–19 Mya has been estimated [25]. This last estimation differs from the age of 35 Mya estimated for Cactaceae based on molecular phylogeny [23]. Accordingly, the members of Cactaceae are relatively young in the evolutionary history of Caryophyllales and, thus, the structures of cpDNA found in *Mammillaria* have evolved in a recent diverging process, as none of the three structures have been reported for other, older members of Caryophyllales.

Our results suggest that the structural reconfigurations of cpDNA within the family Cactaceae have occurred frequently. Particularly, such structural changes have mainly involved genes located at the flanking extremes of the LSC and, secondly, genes located at the SSC. Although deep and strong changes occurred in reconfigurations of the IRs, only few gene rearrangements and loss of genes occurred in the LSC, protecting the large blocks of genes involved in photosynthesis (Figure 2). We conclude that no single type of cpDNA structure characterizes all members of *Mammillaria*; however, the presence of IRs is a marked difference, with respect to the large columnar cacti species that have lost them.

In addition, our results showed that, among the tribes of the subfamily Cactoideae, there are notable structural differences in the cpDNA. The most evident is that *Mammillaria* (Cacteae) has IRs, as occurs in most of Caryophyllales, which were lost in Echinocereeae. We propose that the presence of IRs is the basal state in the Cactaceae family, as the presence of IRs is common in all members of Caryophyllales currently sequenced. However, we cannot ignore the possibility that IRs may be have been lost and recovered (with a new configuration) in multiple evolutionary events occurring during Cactaceae radiation. The phylogenetic relationships of the seven species based on DNA sequences (Figure 4) showed that *M. albiflora* (structure 1) is the closest species to columnar saguaro, however, *M. pectinifera* (structure 1) is the closest one to *M. solisioides* (structure 2). Unexpectedly, *M. zephyranthoides* (structure 3) occupied a branch that is closer to the three species of series *Supertextae*. These phylogenetic results should be taken with caution, since the number of species is too poor; however, it seems that the evolutionary underlying processes that have operated on the structural changes differ of those operating at the level of DNA sequences.

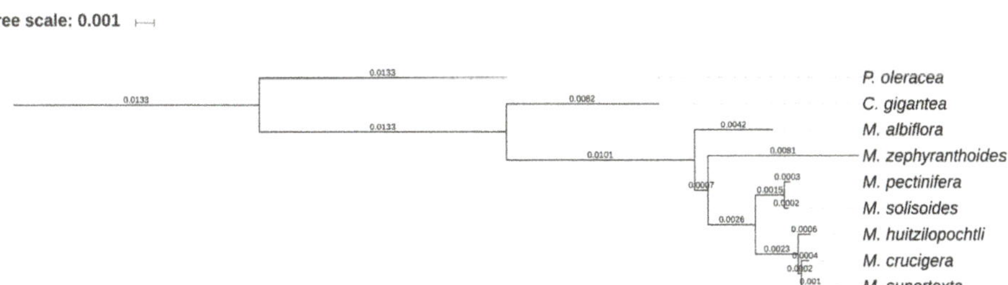

Figure 4. Phylogenetic ML tree obtained for the seven species of *Mammillaria*. The analysis is based on 42 coding regions shared to the two species used as outgroups (*C. gigantea* and *P. oleraceae*).

Our findings showed that, in *Mammillaria*, the concatenated battery *rpl23-trnI-CAU-ycf2* might have a relevant role in the reconfiguration of IRs. Particularly in these genes, it is worthy to highlight the gene *trnI-CAU*, which (along *with trnI-GAU, trnfM-CAU,* and *trnM-CAU*) has been identified, in experimental essays, as one of four essential tRNAs in plastids [26]. In addition, the IRs composed of a single gene have been shown to correspond to *trnI-CAU* (e.g. *Pinus massoniana*) [27]. In this context, we propose that *trnI-CAU* in *Mammillaria* plays a key role in the reconfiguration of IRs, but future studies are needed to verify this.

We found that the IGS *psbK-trnQ* was inserted in both IRs of structure 2, although this IGS at IRB was flanked by *rps19* at the LSC; meanwhile, in the rest of the species, this pair of genes (*psbk* and *trnQ*) was located at the LSC. In addition, the DNA sequence of this IGS was highly conserved, showing 91–97% of identity to the IGS of other Caryophyllales species (e.g., *C. gigantea, Cistanthe longiscaspa, Tallinum paniculatum,* and *P. oleracea*). In addition, we found other highly conserved IGS, which showed 90–98% of identity to the IGS of *trnG-UCC—trnS-GGU* in other members of Caryophyllales. However, in *Mammillaria*, this pair of transfer genes was at the contrary extreme of the LSC, although the DNA sequence of this IGS was found between *rpl20-rps12*. Thus, we support the idea that IGS may have played an important role in the functional processes, in agreement with former studies [28]. Pseudogenization was identified in the seven species of *Mammillaria*, but in structure 3 it was more evident (Figures 1 and 2); particularly by the strong shortening of the genes of the two open reading frames, *ycf2* and *ycf1*, as they had a length of <1 kb whilst, in the other six species of *Mammillaria*, each of these genes had a length of nearly 6–7 kb. The pseudogenization of *ycf1* and *ycf2* indicates a loss of functional activity, which disagrees with the conclusion that these genes are essential for plant survival [29]. In addition, pseudogenization was also identified for all species of *Mammillaria*, with incomplete copies of subunits of *rps* and *rpl* suites, *accD*, as well as, *rps12* and *clpP* duplicated and, evidently, in three dehydrogenase subunits (B, D, and F). These subunits translated to an interrupted sequence of amino acids, which indicates that the functionalities of these genes may have been lost. In addition, seven other subunits of *ndh* genes were completely lost (A, C, E, H, I, J, and K). These subunits in *C. gigantea*, both pseudogenization and the complete loss of *ndh* subunits in cpDNA have also been documented [12]. However, we could not show, for *Mammillaria*, that all of those genes documented as pseudogenes, or entirely lacking in the cpDNA, were not present in nuclear or mitochondrial genomes; in other plants, many genes of the chloroplast have been found in nuclear or mitochondrial genomes [6].

An interesting result, obtained in the SSC of the seven species of *Mammillaria, C. gigantea,* and *P. schottii*, is that they all have identical order and orientation of *ycf2-trnL-CAA-ycf1*. This result suggests that the arrangement of these three genes may be a synapomorphy in the subfamily Cactoideae, as it is not present in *Opuntia microdasys*, subfamily Opuntioideae (sequence consulted GenBank: HQ664651.1), nor in the rest of the species of Caryophyllales [13].

Structure 2 was distinguished by the insertion of the *matK* gene into the IRs, which was nested inside of two complete *trnK* introns. The gene *matK* was also documented in the IRs of some species of *Erodium* (Geraniaceae) but was truncated and separated to the two *trnK* introns [17]. In addition, in IRs of *Lamprocapnos spectabilis* (Papaveraceae), complete *matK* nested between the two introns has been documented [30]. It is relevant to point out that *matK* is duplicated in these genomes due to its insertion in IRs. This encoding gene plays a fundamental role in photosynthesis by editing the RNAs of nearly 15 proteins, even though the *trnK* introns were lost [31]. Thus, in species having structure 2, by the formation of IRs, there are two copies of *matK* in the haploid genome of the chloroplast; however, we do not possess sufficient information to discuss the functional consequences of this.

Finally, it is important to point out the following: 1) the three structures of cpDNA of *Mammillaria* are not concordant to the taxonomic subgenera and series levels. Consequently, our results do not support any of the infrageneric classification currently proposed. 2) *Sensu* [18], *M. solisioides* is considered to be a subspecies of *M. pectinifera* (subgenus *Krainzia*, series *Herrerae+Pectiniferae*); however, *M. solisioides* exhibits the cpDNA structure found in the *Supertextae*, subgenus *Mammillaria*. 3)

Accordingly, we consider that *M. solisioides* is an independent species, in agreement with Arias et al., [32]. We conclude that the structural analysis of cpDNA can also contribute towards clarifying the taxonomic relationships of *Mammillaria* with other plant species.

4. Materials and Methods

4.1. Plant Sampling and DNA Extraction

We included seven species of *Mammillaria*, which represent three of the eight subgenera proposed by Hunt [18]. These seven species are listed in the IUCN red list [19]. The three species sampled (*M. crucigera, M. huitzilopochtli,* and *M. supertexta*) are classified in the subgenus *Mammillaria*, series *Supertextae*. According to Crozier [21], only the species included in this subgenus represent a natural and monophyletic clade in phylogenies obtained with chloroplast DNA sequences. The other three taxa sampled (*M. albiflora, M. pectinifera,* and *M. solisioides)* are of the subgenus *Krainzia*, series *Herrerae+Pectiniferae*. It is important to mention that *M. solisioides* was considered [18] to be a subspecies of *M. pectinifera*. The last sampled species was *M. zephyranthoides*, classified in the subgenus *Dolichothele*. This last species has faced a complex and controversial taxonomic identification, as it has been included in both *Dolichothele* and *Mammillaria*.

Living complete plants for tissue samples were obtained for *M. albiflora, M. crucigera, M. huitzilopochtli, M. pectinifera, M. solisioides,* and *M. supertexta*. Small pieces of the surface of green stems were cleaned of spines and areoles to obtain 300 mg samples of tissue, which were treated using a MinuteTM Chloroplast Isolation Kit (Invent Technologies Inc., Eden Prairie, MN, USA), according to the manufacturer's instructions. The chloroplast extracts were processed with a DNeasy plant minikit (Qiagen, Valencia, CA, USA), in order to obtain enriched chloroplast genomes.

4.2. High-Throughput Sequencing and Sanger Verification

Massive sequencing was done with the Illumina platform (Ilumina, San Diego, CA, USA). For each species, genomic libraries were prepared with the Nextera XT kit, according to the manufacturer's instructions, and sequenced in a MiSeq 2 × 300 cycles. The flanks of IRs and gene rearrangements were PCR verified (Table S1), using the recently assembled cpDNA in our study for the design of specific primers with Primer3 v.4 [33]. These PCR products were sequenced in a 3730×l capillary sequencer (Applied Biosystems, Pleasanton, CA, USA).

4.3. Genome Assembly, Annotation, and Structural Alignment

The available assembly genome for the giant columnar cactus species of *Carnegiea gigantea* [12] lacks IRs and, thus, de novo assembly was carried out with NovoPlasty v.2.6.5. [34] and DISCOVAR de novo v.52488 [35]. The scaffolding was carried out with Ragout v.2.0 [36]. The gaps were filled with GARM v.0.7.5 [37] and the circularization of each of the assembly genomes was obtained with Circlator v.1.5.5 [38]. The assemblage of each genome was tested with REAPR v.1.0.18 [39]. The annotation for the seven species was done with GeSeq [40] and the genomes were drawn with OGD [41]. For gene annotation, we used the cpDNA of *C. gigantea* [12] and *Portulaca oleracea* [42]. Annotation of the largest genome found for each of the three different structures (Table 1) was manually curated. The structural alignment of the complete assembly genomes was performed with MAUVE [43]. This program was also used to compare these structures to other species of Caryophyllales (*P. oleracea*, NC_036236.1) and Cactaceae (*C. gigantea*, GCA_002740515.1), in order to emphasize the relevance of our structural findings in the whole chloroplast genome of *Mammillaria*. In order to reconstruct the phylogenetic relationships of the seven species of *Mammillaria*, two species (*C. gigantea* and *P. oleracea*) were used as outgroups. A total of 42 orthologous protein-coding genes (Table S2) shared in these nine species were identified with Prottest [44]. The DNA sequences of these 42 loci were aligned using MAFFT v7.310 [45]. The Akaike Information Criterion (AIC) in JMODELTEST v2.1.10 was used to determine the best-fitting model of nucleotide substitutions [46]. The GTR + G model was used to obtain the phylogenetic tree based on ML in RAXML-HPC v8.2.10 [47] with 1000 replicates.

Supplementary Materials: The following are available online at http://www.mdpi.com/2223-7747/8/10/392/s1, Table S1: Primer sequences designed for PCR verifications for flanking adjacent sequences of IRs; and gene rearrangements located at LSC in three genomes structures of *Mammillaria*. In the last column the temperature used to amplify each locus for all species. Primer design was based on the DNA sequences of *M. albiflora* (structure 1), *M. supertexta* (structure 2) and *M. zephyranthoides* (structure 3); Table S2. List of the total of 42 coding loci were used to obtain phylogenetic relationships of the seven species of *Mammillaria* studied and the two species used as outgroups *Carnegiea gigantea and Portulaca oleraceae.*

Author Contributions: S.S. is the researcher leading of this study, she designed research and obtain the financial support. S.S., S.A., and P.D. carried out fieldwork and taxonomic identification. S.S. and D.A.C. carried out chloroplasts and DNA isolation. C.E.D.-V. and F.V.-P. implemented experimental protocols of massive sequencing. A.S.-F. supervised bioinformatics analyses, which were processed by D.A.C. and K.E. A.G.-R. supervised the experiments for PCR verifications. S.S. and D.A.C. prepared the complete manuscript; and A.G.-R. and P.D. reviewed all draft versions. All the authors approved the submitted version of this this article.

Funding: This study is supported by grants from the Dirección General de Personal Académico of UNAM for research projects (DGAPA-PAPIIT IN222216).

Acknowledgments: SEMARNAT (SGPA/DGVS/06880/16) supplied sampling permission. G. Mendoza-Juárez (IIES UNAM) carried out PCR verification. L.M Marquez-Valdelamar and N.M López-Ortiz (Laboratorio de Secuenciación de LaNaBio, IB UNAM) provided sequencing service for Sanger sequencing. P. Gaytán (Unidad de Síntesis y Secuenciación de ADN, IBT UNAM) synthetized primers for PCR verification. We are grateful to Unidad de Secuenciación Masiva y Bioinformática of the Laboratorio Nacional de Apoyo Tecnológico a las Ciencias Genómicas, CONACyT #260481, at the Instituto de Biotecnología/UNAM for advice and training in bioinformatics to Sofía Solórzano and Delil A. Chincoya. The comments of two anonymous reviewers improved the quality of this article.

Conflicts of Interest: The authors declare that the research was conducted in the absence of any commercial or financial relationships that could be construed as a potential conflict of interest.

References

1. Daniell, H.; Lin, C.S.; Yu, M.; Chang, W.J. Chloroplast genomes: Diversity, evolution, and applications in genetic engineering. *Genome Biol.* **2016**, *17*, 134. [CrossRef] [PubMed]
2. Morden, C.; Delwiche, C.; Kuhsel, M.; Palmer, D. Gene phylogenies and the endosymbiotic origin of plastids. *Biosystems* **1992**, *28*, 75–90. [CrossRef]
3. Cavalier-Smith, T. The origins of plastids. *Biol. J. Linn. Soc.* **1982**, *17*, 289–306. [CrossRef]
4. Lemieux, C.; Otis, C.; Turnel, M. Ancestral chloroplast genome in *Mesostigma viride* reveals an early branch of green plant evolution. *Nature* **2000**, *403*, 649–652. [CrossRef] [PubMed]
5. Xiao-Ming, Z.; Junrui, W.; Li, F.; Sha, L.; Hongbo, P.; Lan, Q.; Jing, L.; Yan, S.; Weihua, Q.; Lifang, Z.; et al. Inferring the evolutionary mechanism of the chloroplast genome size by comparing whole-chloroplast genome sequences in seed plants. *Sci. Rep.* **2017**, *7*, 1555. [CrossRef] [PubMed]
6. Thorsness, P.E.; Weber, E.R. Escape and migration of nucleic acids between chloroplasts, mitochondria, and the nucleus. *Int. Rev. Cytol.* **1996**, *165*, 207–234. [CrossRef]
7. Yagi, Y.; Shiina, T. Recent advances in the study of chloroplast gene expression and its evolution. *Front. Plant Sci.* **2014**, *5*, 61. [CrossRef]
8. Kung, S.D.; Lin, C.M. Chloroplast promoters from higher plants. *Nucleic Acids Res.* **1985**, *13*, 7543–7549. [CrossRef]
9. Mower, J.P.; Vickrey, T.L. *Advances in Botanical Research Plastid Genome Evolution*; Chaw, S.M., Jansen, R.K., Eds.; Academic Press of Elsevier: London, UK, 2018; Volume 85, Chapter 9; pp. 263–292.
10. Zhu, A.; Guo, W.; Sakski, G.; Weishu, F.; Mover, J.P. Evolutionary dynamics of the plastid inverted repeat: The effects of expansion, contraction, and loss of substitution rates. *New Phytol.* **2016**, *209*, 1747–1756. [CrossRef]
11. Lavin, M.; Doyle, J.J.; Palmer, J.D. Evolutionary significance of the loss of the chloroplast-DNA Inverted Repeat in the Leguminosae Subfamily Papilionoideae. *Evolution* **1990**, *44*, 390–402. [CrossRef]
12. Sanderson, M.J.; Copetti, D.; Búrquez, A.; Bustamante, E.; Charboneau, J.L.M.; Eguiarte, L.; Kumar, S.; Lee, H.O.; McMahon, M.; Steele, K.; et al. Exceptional reduction of the plastid genome of saguaro cactus (*Carnegiea gigantea*). *Am. J. Bot.* **2015**, *102*, 1115–1127. [CrossRef] [PubMed]

13. Yao, G.; Jin, J.J.; Li, H.T.; Yang, J.B.; Mandala, V.S.; Croley, M.; Mostow, R.; Douglas, N.A.; Chase, M.W.; Christenhusz, M.J.M.; et al. Plastid phylogenomic insights into the evolution of Caryophyllales. *Mol. Phylogenet. Evol.* **2019**, *134*, 74–86. [CrossRef] [PubMed]

14. Huang, H.; Shi, C.; Liu, Y.; Mao, S.Y.; Gao, L.Z. Thirteen *Camellia* chloroplast genome sequences determined by high-throughput sequencing: Genome structure and phylogenetic relationships. *BMC Evol. Biol.* **2014**, *14*, 151. [CrossRef] [PubMed]

15. Wu, Z.; Tembrock, L.R. Two complete chloroplast genomes of white campion (*Silene latifolia*) from male and female individuals. *Mitochondrial DNA Part A* **2015**, *28*, 375–376. [CrossRef]

16. Kang, J.S.; Lee, B.Y.; Kwak, M. The complete chloroplast genome sequences of *Lychnis wilfordii* and *Silene capitata* and comparative analyses with other Caryophyllaceae genomes. *PLoS ONE* **2017**, *27*, e0172924. [CrossRef] [PubMed]

17. Blazier, J.C.; Jansen, R.K.; Mower, J.P.; Govindu, M.; Zhang, J.; Weng, M.L.; Ruhlman, T.A. Variable presence of the inverted repeat and plastome stability in *Erodium. Ann. Bot.* **2016**, *117*, 1209–1220. [CrossRef]

18. Hunt, D.; Taylor, N.; Charles, G. *The New Cactus Lexicon*; DH Books: Milborne Port, UK, 2006.

19. IUCN International Union for Conservation of Nature. Consulted for Mammillaria genus. Available online: https://www.iucnredlist.org/search/list?query=Mammillaria&searchType=species (accessed on 30 June 2019).

20. Butterworth, C.A.; Wallace, R.S. Phylogenetic studies of *Mammillaria* (Cactaceae)—Insights from chloroplast sequence variation and hypothesis testing using the parametric bootstrap. *Am. J. Bot.* **2004**, *91*, 1086–1098. [CrossRef] [PubMed]

21. Crozier, B.S. Systematics of Cactaceae Juss: Phylogeny, cpDNA Evolution, and Classification, with Emphasis on the Genus *Mammillaria* Haw. Ph.D. Thesis, University of Texas, Austin, TX, USA, 2005.

22. Hernández-Hernández, T.; Hernández, H.M.; De Nova, J.A.; Puente, R.; Eguiarte, L.; Magallón, S. Phylogenetic relationships and evolution of growth form in Cactaceae (Caryophyllales, Eudicotyledoneae). *Am. J. Bot.* **2011**, *98*, 44–61. [CrossRef]

23. Arakaki, M.; Christin, P.A.; Nyffeler, R.; Lendel, A.; Eggli, U.; Ogburn, M.; Spriggs, E.; Moore, M.J.; Edwards, E.J. Contemporaneous and recent radiations of the world's major succulent plant lineages. *Proc. Natl. Acad. Sci. USA* **2011**, *108*, 8379–8384. [CrossRef]

24. Turmel, M.; Otis, C.; Lemieux, C. Divergent copies of the large inverted repeat in the chloroplast genomes of ulvophycean green algae. *Sci. Rep.* **2017**, *7*, 994. [CrossRef]

25. Ocampo, G.; Columbus, J.T. Molecular phylogenetics of suborder Cactineae (Caryophyllales), including insight into photosynthetic diversification and historical biogeography. *Am. J. Bot.* **2010**, *97*, 1827–1847. [CrossRef] [PubMed]

26. Alkatib, S.; Fleischmann, T.T.; Scharff, L.B.; Bock, R. Evolutionary constraints on the plastid tRNA set decoding methionine and isoleucine. *Nucleic Acid Res.* **2012**, *40*, 6713–6724. [CrossRef] [PubMed]

27. Ni, Z.; Ye, Y.; Bal, T.; Xu, M.; Xu, L.A. Complete chloroplast genome of *Pinus massoniana* (Pinaceae): Gene rearrangements, loss of ndh genes, and short inverted repeats contraction, expansion. *Molecules* **2017**, *22*, 1528. [CrossRef]

28. Hao, D.C.; Chen, S.L.; Huang, B.L. Evolution of the chloroplast trnL-trnF region in the Gymnosperm lineages Taxaceae and Cephalotaxaceae. *Biochem. Genet.* **2009**, *47*, 351–369. [CrossRef] [PubMed]

29. Drescher, A.; Ruf, S.; Calsa, T., Jr.; Carrer, H.; Bock, R. The two largest chloroplast genome-encoded open reading frames of higher plants are essential genes. *Plant J.* **2000**, *22*, 97–104. [CrossRef] [PubMed]

30. Park, S.; An, B.; Park, S. Reconfiguration of the plastid genome in *Lamprocapnos spectabilis*: IR boundary shifting, inversion, and intraspecific variation. *Sci. Rep.* **2018**, *8*, 13568. [CrossRef] [PubMed]

31. Barthet, M.M.; Hilu, K.W. Expression of *matK*: Functional and evolutionary implications. *Am. J. Bot.* **2007**, *94*, 1402–1412. [CrossRef]

32. Arias, S.; Gama-López, S.; Guzmán-Cruz, L.U.; Vázquez-Benítez, B. Cactaceae. In *Flora del Valle de Tehuacán-Cuicatlán*; Medina, L.R., Ed.; Instituto de Biología, Universidad Nacional Autónoma de México: Mexico City, Mexico, 2012; Volume 95.

33. Rozen, S.; Skaletsky, H. Primer3 on the WWW for general users and for biologist programmers. *Methods Mol. Biol.* **2000**, *132*, 365–386. [CrossRef]

34. Dierckxsens, N.; Mardulyn, P.; Smits, G. NOVOPlasty: De novo assembly of organelle genomes from whole genome data. *Nucleic Acids Res.* **2017**, *45*, e18. [CrossRef]

35. Love, R.R.; Weisenfeld, N.I.; Jaffe, D.B.; Besansky, N.J.; Neafsey, D.E. Evaluation of DISCOVAR de novo using a mosquito simple for cost-effective short-read genome assembly. *BMC Genom.* **2016**, *17*, 18. [CrossRef]

36. Kolmogorov, M.; Raney, B.; Paten, B.; Pham, S. Ragout-a reference-assisted assembly tool for bacterial genomes. *Bioinformatics* **2014**, *30*, i302–i309. [CrossRef] [PubMed]

37. Soto-Jiménez, L.M.; Estrada, K.; Sanchez-Flores, A. GARM: Genome assembly, reconciliation and merging pipeline. *Curr. Top. Med. Chem.* **2014**, *14*, 418–424. [CrossRef] [PubMed]

38. Hunt, M.; De Sila, N.; Otto, T.D.; Parkhill, J.; Keane, J.A.; Harris, S.R. Circlator: Automated circularization of genome assemblies using long sequencing reads. *Genome Biol.* **2015**, *16*, 294. [CrossRef] [PubMed]

39. Hunt, M.; Kikuchi, T.; Sanders, M.; Newbold, C.; Berriman, M.; Otto, T.D. REAPR: A universal tool for genome assembly evaluation. *Genome Biol.* **2013**, *14*, R47. [CrossRef] [PubMed]

40. Tillich, M.; Lehwark, P.; Pellizzer, T.; Ulbricht-Jones, E.S.; Fischer, A.; Bock, R.; Greiner, S. GeSeq–versatile and accurate annotation of organelle genomes. *Nucleic Acids Res.* **2017**, *45*, W6–W11. [CrossRef] [PubMed]

41. Lohse, M.; Drechsel, O.; Bock, R. OrganellarGenomeDRAW (OGDRAW): A tool for the easy generation of high-quality custom graphical maps of plastid and mitochondrial genomes. *Curr. Genet.* **2007**, *52*, 267–274. [CrossRef]

42. Liu, X.; Yang, H.; Zhao, J.; Zhou, B.; Li, T.; Xiang, B. The complete chloroplast genome sequence of the folk medicinal and vegetable plant purslane (*Portulaca oleracea* L.). *J. Hortic. Sci. Biotech.* **2018**, *93*, 356–365. [CrossRef]

43. Darling, A.C.; Mau, B.; Blattner, F.R.; Perna, N.T. Mauve: Multiple alignment of conserved genomic sequence with rearrangements. *Genome Res.* **2004**, *14*, 1394–1403. [CrossRef]

44. Darriba, D.; Taboada, G.L.; Doallo, R.; Posada, D. ProtTest 3: Fast selection of best-fit models of protein evolution. *Bioinformatics* **2011**, *27*, 1164–1165. [CrossRef]

45. Katoh, K.; Daron, M.S. MAFFT multiple sequence alignment software version 7: Improvements in performance and usability. *Mol. Biol. Evol.* **2013**, *30*, 772–780. [CrossRef]

46. Posada, D. JModelTest: Phylogenetic model averaging. *Mol. Biol. Evol.* **2008**, *25*, 1253–1256. [CrossRef] [PubMed]

47. Stamatakis, A. RAxML version 8: A tool for phylogenetic analysis and post-analysis of large phylogenies. *Bioinformatics* **2014**, *30*, 1312–1313. [CrossRef] [PubMed]

 *plants*

Article

New Insights on *Lilium* Phylogeny Based on a Comparative Phylogenomic Study Using Complete Plastome Sequences

Hyoung Tae Kim [1], Ki-Byung Lim [2] and Jung Sung Kim [3,*]

[1] Institute of Agricultural Science and Technology, Chungbuk National University, Chungbuk, 28644, Korea; rladbgus@gmail.com

[2] Department of Horticulture, Kyungpook National University, Daegu, 41566, Korea; kblim@knu.ac.kr

[3] Department of Forest Science, Chungbuk National University, Chungbuk, 28644, Korea

* Correspondence: jungsung@chungbuk.ac.kr; Tel.: +82-43-261-2535

Received: 6 November 2019; Accepted: 22 November 2019; Published: 27 November 2019

Abstract: The genus *Lilium* L. is widely distributed in the cold and temperate regions of the Northern Hemisphere and is one of the most valuable plant groups in the world. Regarding the classification of the genus *Lilium*, Comber's sectional classification, based on the natural characteristics, has been primarily used to recognize species and circumscribe the sections within the genus. Although molecular phylogenetic approaches have been attempted using different markers to elucidate their phylogenetic relationships, there still are unresolved clades within the genus. In this study, we constructed the species tree for the genus using 28 *Lilium* species plastomes, including three currently determined species (*L. candidum*, *L. formosanum*, and *L. leichtlinii* var. *maximowiczii*). We also sought to verify Comber's classification and to evaluate all loci for phylogenetic molecular markers. Based on the results, the genus was divided into two major lineages, group A and B, consisting of eastern Asia + Europe species and Hengduan Mountains + North America species, respectively. Sectional relationships revealed that the ancestor *Martagon* diverged from *Sinomartagon* species and that *Pseudolirium* and *Leucolirion* are polyphyletic. Out of all loci in that *Lilium* plastome, *ycf1*, *trnF-ndhJ*, and *trnT-psbD* regions are suggested as evaluated markers with high coincidence with the species tree. We also discussed the biogeographical diversification and long-distance dispersal event of the genus.

Keywords: *Lilium*; phylogenomics; plastome; molecular markers; gene tree; species tree

---

1. Introduction

The genus *Lilium* L. is the type genus of Liliaceae that consists of approximately 100 species spread throughout the cold and temperate regions of the Northern Hemisphere [1,2]. *Lilium* species are economically important because of their ornamental features in horticulture as cut flowers and as potted and garden plants [3]. In addition, the flowers and bulbs of cultivated *Lilium* species are used for food and medicine [4].

The classification of the genus *Lilium* had been built based on flower shape before Comber's classification [5]. As a result, the sectional (or subgeneral) boundaries in the genus frequently changed, and many species were referred to different sections according to different classification systems [6–8]. In contrast to the previous sectional concept based on flower shape, Comber [5] suggested the use of new natural characteristics, including characteristics of leaves, bulbs, and stems to classify the genus *Lilium* and divided it into seven sections. Although Comber's classification was revised by De Jong [9] based on previous papers published up to that time, Comber's classification has been primarily used to date to recognize *Lilium* species and circumscribe the sections within the phylogeny of *Lilium*.

Nishikawa et al. [10] constructed the phylogeny of 55 *Lilium* species using the internal transcribed spacer (ITS) region to clarify their phylogenetic relationships and found that most species formed a clade according to the classification based on the morphological features. Nonetheless, it is difficult to follow their results directly because most of the branch lengths were very short and not significantly supported. In addition, the phylogenies using the same marker with more samples indicated that most of the sections, except for *Martagon*, were polyphyletic [11–13]. Such discordance between classifications based on morphological characteristics and molecular phylogeny has also been found in other genera. For example, the subgenera of the genus *Cymbidium* based on morphological characteristics [14] was not found to be monophyletic using ITS [15] and ITS + *matK* [16,17], and a number of branches in the phylogenetic trees were collapsed in strict consensus trees. It is likely that gene trees based on molecular markers were often insufficient for investigating species trees because of certain evolutionary events, such as incomplete lineage sorting and horizontal gene transfer, or because convergent morphological evolution has occurred in certain lineages. To ensure the phylogenetic accuracy, increased taxa sampling has been preferred for a long time [18–20]. However, it has been suggested that the number of genes may be a more important determinant than the taxon number [21–23], and the rapid development of next-generation sequencing techniques has led us to the phylogenomic era, which provides numerous data to resolve ambiguous relationships.

Plants contain three genomes: a nuclear genome, a mitochondrial genome, and a plastid genome (plastome). The nuclear genomes [24] and mitochondrial genomes [23] in flowering plants substantially vary in length, whereas the plastomes maintain consistent lengths and typical structures for a long time [25], except for those of some specific lineages [26–28]. In addition, the typical plastome structure allows for next-generation sequencing (NGS) data to be assembled easily. As a result, the number of deposited plastome sequences is 10 times that of mitochondrial genome sequences in the NCBI genome database. The highly conservative nature of the plastome structure makes it possible for intergenic spaces to be used as molecular markers, as well as genes [29,30]. Using this feature, many researchers have attempted to increase phylogenetic accuracy for unresolved taxa using whole plastome sequences [31–33] and have also suggested new combinations of molecular markers for specific taxa [34–38]. Two studies on the phylogenomics of *Lilium* using plastome sequences have been published [4,39] to date. The phylogenetic relationships among the branches were well resolved with high support. However, taxon samplings were restricted because of the inability to use the data from both papers, which were published in the same year. Therefore, it was difficult to verify Comber's classification based on well-resolved phylogeny. Du et al. [4] suggested molecular markers for the phylogeny of *Lilium* based on nucleotide diversity; however, the issues of polytomies and low supporting values still remained. Apparently, there was no comparison between species trees and gene trees to suggest molecular markers for the phylogenetic analysis. A gene tree can be easily acquired from each gene; however, it can conflict with the species tree [40] because not all genes have evolved in the same manner in the same lineage. Therefore, it is necessary to evaluate the reliability of any constructed species trees and gene trees and suggest molecular markers for the phylogeny of certain lineages.

In this study, three plastomes in *Lilium* were newly sequenced (1) to verify Comber's classification and (2) to evaluate all loci of the *Lilium* plastome for phylogenetic molecular markers. Two taxa, *Lilium formosanum* Wallace and *Lilium candidum* L., were sampled to investigate the monophyly of *Leucolirion* and to determine the phylogenetic position of *Liriotypus*. *Lilium leichtlinii* var. *maximowiczii* (Regel) Baker was added because it was found to be closely or distantly related to *Lilium lancifolium* Thunb. in the previous studies [10–13]. The phylogenies of the genus *Lilium* that were built based on different methodologies using plastome sequences were constructed to determine a more accurate species tree. Based on this result, gene trees were compared to the species tree of the genus *Lilium* to evaluate which gene trees were similar to the species tree in a topology, while reflecting the generic relationships in Liliaceae. Additionally, based on the species tree generated in this study, the evolution of genomic characteristics in the genus *Lilium* was also discussed.

2. Results

2.1. Newly Sequenced Plastomes of Three Lilium Species

Numbers of reads ranging from 35,028,408 (*L. leichtlinii* var. *maximowiczii*) to 89,224,524 (*L. candidum*) were used for raw data after removing less than 50 bp reads from each dataset for the three species (Table 1). After constructing a complete plastome sequence, the number of mapped reads to complete the plastome sequences ranged from 568,878 (*L. leichtlinii* var. *maximowiczii*) to 1,945,534 (*L. formosanum*). Consequently, the average coverage of each plastome sequence ranged from 520.1 (*L. leichtlinii* var. *maximowiczii*) to 1840.6 (*L. formosanum*). The plastomes of the three *Lilium* species (*L. candidum*, *L. formosanum*, and *L. leichtlinii* var. *maximowiczii*) were 152,101–152,653 bp in length with a large single copy (81,481–82,101 bp), a small single copy (17,524–17,644 bp), and two inverted repeats (26,488–26,514 bp) (Table S1).

Table 1. Summary of genome assembly.

Taxon	No. of Raw Reads (≥50 bp)	No. of Mapped Reads	Average Coverage	SRA [a] Accession
Lilium candidum	89,224,524	1,187,028	1134.7	SRR7617960, SRR7617961
Lilium leichtlinii var. *maximowiczii*	35,028,408	568,878	520.1	SRR7617965
Lilium formosanum	80,985,606	1,945,534	1840.6	SRR7617963, SRR7617964

[a] Sequence Read Archive.

The overall GC content was 37.0%, which is similar to those of other *Lilium* species. In total, 133 genes were annotated from each plastome with 85 protein-coding genes, 8 rRNA genes, 38 tRNA genes, and 2 partial genes (*rps19* and *ycf1*). *InfA* was a pseudogene in all three plastomes, as well as other *Lilium* species, and *cemA* was pseudogenized in the plastomes of *L. candidum* and *L. leichtlinii* var. *maximowiczii* because of the copy number variation of the poly A-tract.

2.2. Nucleotide Diversity within Genera Lilium and Fritillaria

The total length of the aligned sequences of *Lilium* and *Fritillaria* plastomes was 159,458 bp. The maximum nucleotide diversities of *Lilium* and *Fritillaria* were 0.030 and 0.041, respectively (Figure 1). In total, the nucleotide diversity was lower in the inverted repeat (IR) region than in the large single copy (LSC) and small single copy (SSC) regions in both genera, as well as being inversely proportional to the GC content. The delta nucleotide diversity between the two genera fluctuated from 0.014 to −0.017. Four loci higher than 0.02 in *Lilium* plastomes and two loci higher than 0.025 in *Fritillaria* plastomes were caused by large deletions in certain species because we deleted the sites with missing data for at least one species. In addition, there were the loci with higher nucleotide diversity, particularly in *Lilium* plastome sequences, which were caused by small palindromic repeats and the IR expansion from IRa to SSC.

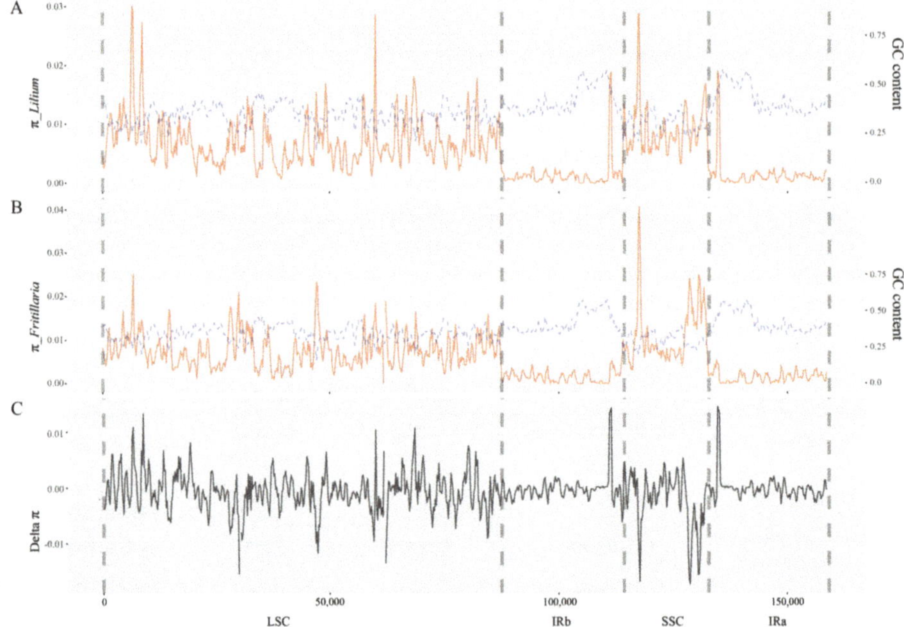

Figure 1. Nucleotide diversity (π) and GC content throughout the plastome sequence according to sliding window analysis (window size = 600 bp, step size = 200 bp). The red line and blue dashed line refer to π and GC content, respectively. The vertical dashed grey lines refer to the approximate boundaries of the plastome structure.

2.3. Phylogeny of Lilium in Liliaceae

The 28 phylogenetic trees constructed from the four datasets, four tools, and two models showed that all of the genera in Liliaceae were monophyletic (Figure 2). *Amana* and *Erythronium* in tribe Tulipeae were distinguished from *Cardiocrinum*, *Fritillaria*, and *Lilium* in tribe Lilieae. In tribe Lilieae, *Cardiocrinum* diverged first, and *Lilium* and *Fritillaria* were separated later. In contrast to the generic relationships within the tribe, the infrageneric relationships varied slightly among phylogenetic trees.

In the clades of *Lilium* in 28 phylogenies, different datasets affected the change in topology more than different tools and models; however, the major clades were highly conserved in all phylogenies (Figure S1). The genus was divided into two major lineages, group A and group B, consisting of eastern Asia + Europe species and Hengduan Mountains + North America species, respectively (Figure 3). Group A consisted of three clades and five independent lineages. Clade I comprised three *Martagon* species and *L.* sp. KHK_2014, except for the phylogeny created by ASTRAL with coding genes. Clade II consisted of *L. amabile*, *L. lancifolium*, and *L. callosum*, which belong to *Sinomartagon*. *L. longiflorum*, and *L. formosanum* of *Leucolirion* and *L. brownii* of *Archelirion* formed clade III. *L. cernuum* of *Sinomartagon* and *L. candidum* of *Liriotypus* were sister groups to *Martagon* and the rest of group A, respectively. The positions of *L. leichtlinii* var. *maximowiczii* were distantly related to *L. lancifolium* and *L. amabile*, which were morphologically close to *L. leichtlinii* var. *maximowiczii*.

Group B consisted of four clades and two independent lineages. Clade IV comprised species in *Sinomartagon* 5c. Clade V consisted of *L. leucanthum* of *Leucolirion* and *L. henryi* of *Sinomartagon* 5a. Clade VI comprised *L. duchartrei* and *L. fargesii* of *Sinomartagon*, and clade VII included *L. washingtonianum*, *L. pardalinum*, and *L. superbum* of *Pseudolirium*. *L. distichum* of *Martagon* and *L. philadelphicum* of *Pseudolirium* were sisters to clade IV and clade IV + V, respectively. Most of the branches, including seven clades in groups A and B, were strongly supported by bootstrap values of the maximum likelihood analysis, branch support values of ASTRAL, and posterior probabilities of Bayesian inference, but certain branches within clades had moderate support.

A consensus tree of 28 phylogenies was constructed using the 50% majority rule to infer a robust species tree of *Lilium* (Figure 2). Except for two polytomies, one in *Lilium* and another in *Amana*, the seven clades in *Lilium*, as described above, were maintained.

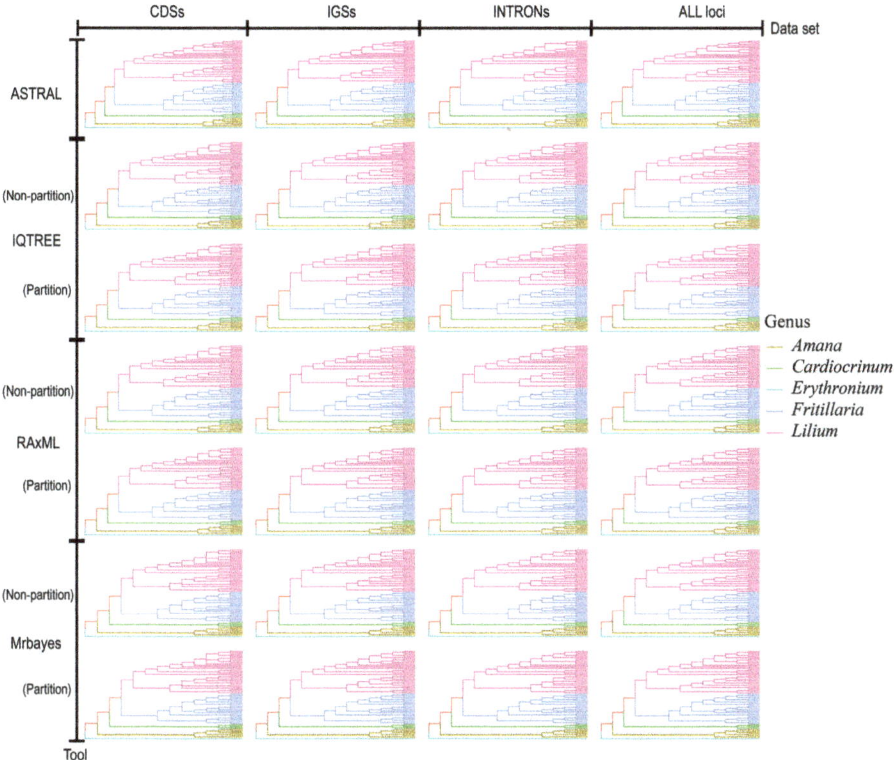

Figure 2. Phylogenetic trees based on four tools (ASTRAL, IQ-TREE, RAxML, and MrBayes) using four datasets (genes, introns, intergenic spacers, and all regions) and two different models (partition model and non-partition model of the dataset). The colored line refers to each genus.

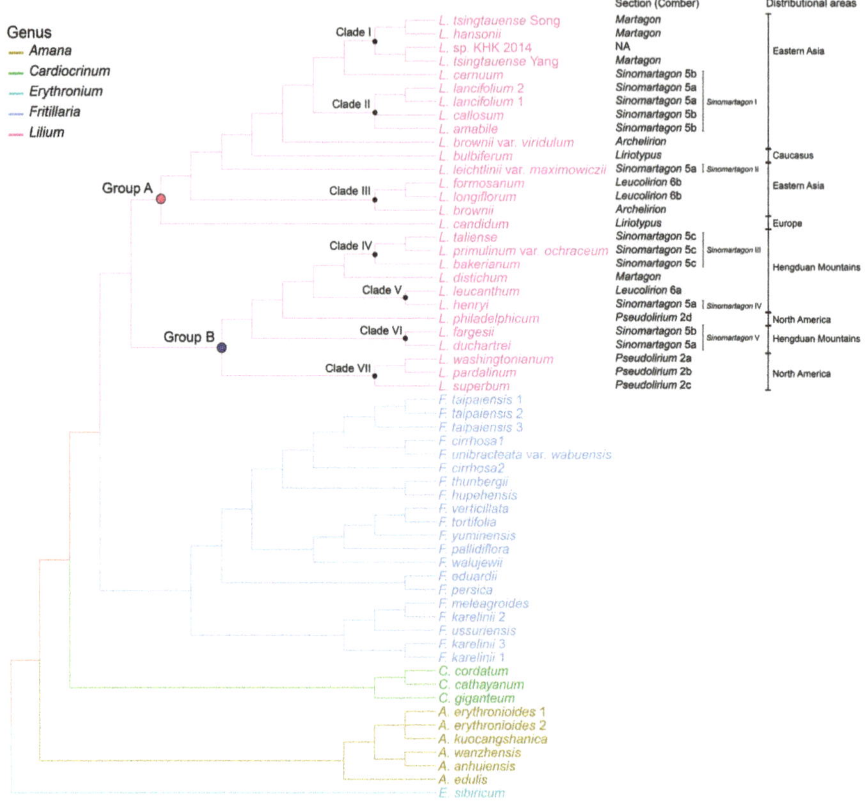

Figure 3. Consensus tree of 28 phylogenies based on four tools (ASTRAL, IQ-TREE, RAxML, and MrBayes) using four datasets (genes, introns, intergenic spacers, and all regions) and two different models (partition model and non-partition model of the dataset). The colored line refers to each genus. The distribution areas of clades are based on Gao et al. [41] and Xinqi et al. [42].

2.4. Comparing Gene Trees to Species Trees

Among the 189 total gene trees, only 22 gene trees showed that each genus was monophyletic, even though two trees built using *trnS-trnG* IGS and the *trnV* intron had different generic relationships compared with the other 20 gene trees (Figures S2–S5). Thirty-six clusters of similar trees from the 190 trees (189 gene trees + the consensus tree of 28 species trees) were generated using treespace [43], with five principal components and a cut-off height of 100 (Figure 4). Among them, 21 clusters consisted of more than three trees (Table 2), and the 10th cluster included a consensus tree of the 28 species trees and 21 gene trees. These 21 gene trees were identical to the gene trees wherein each genus was monophyletic, except for the *ycf2* gene tree.

Table 2. Information on 21 clustered trees using treespace, with five principal components and 100 cut-off distance.

Cluster	Clustered Trees	No. of Trees
1	*accD; ndhG; psbB; rpl20; rpl32; rpl33; rps19; accD_psaI; psaJ_rpl33; trnS_trnG; ycf4_cemA; trnK_intron2*	12
2	*atpA; ccsA; ndhA; ndhH; atpF_atpH; petA_psbJ; rpoB_trnC; rps15_ycf1; rps16_trnQ; trnE_trnT; trnT_trnL; rpoC1_intron*	12
3	*atpE; rps15; ndhH_rps15; ndhJ_ndhK; trnW_trnP*	5
4	*atpF; petG; psbH; psbT; ndhG_ndhI; petB_petD; psbA_trnK; trnN_ycf1*	8
5	*atpH; ndhC; psaI; psbE; psbF; psbJ; rpl2; rpl16; rpl23; ccsA_ndhD; petL_petG; trnA_rrn23; trnV_rrn16*	13
6	*atpI; ndhD_psaC; ndhF_rpl32; rpl14_rpl16; rps19_trnH; rrn5_trnR; trnS_rps4*	7
7	*cemA; petB; rbcL; rpl22; rps3; rps18; ndhC_trnV; rps11_rpl36; rps14_psaB; trnK_rps16; trnR_trnN*	11
8	*clpP; rps11; trnI_intron*	3
9	*matK; ndhB; ndhI; rpl14; rps7; rps14; psbE_petL; psbH_petB; psbJ_psbL; rps4_trnT; trnG_trnfM; trnL_ndhB; trnL_trnF; clpP_intron1; petB_intron; trnG_intron*	16
10	*ndhD; ndhF; ndhK; petA; psaA; rpoB; rpoC1; rpoC2; ycf1; ycf2; atpH_atpI; trnF_ndhJ; trnQ_psbK; trnS_psbZ; trnT_psbD; ycf3_trnS; atpF_intron; ndhA_intron; petD_intron; rpl16_intron; rps16_intron;* Consensus tree of species trees	22
11	*ndhE; ndhJ; psbC; rpoA; rps16; clpP_psbB; rps2_rpoC2; ycf2_trnL*	8
12	*petN; psbL; psbN; rps12_intron*	4
13	*psaB; psbA; rps8; psbM_trnD; rrn16_trnI; trnP_psaJ*	6
14	*psaC; psaJ; psbM; ndhB_rps7; psbN_psbH; rpl23_trnI*	6
15	*psbD; psaA_ycf3; rps12_trnV; trnV_trnM; ndhB_intron; rpl2_intron*	6
16	*psbK; psbZ; rps12; psbC_trnS; rrn4.5_rrn5; trnA_intron*	6
17	*rps2; rps4; trnV_intron*	3
18	*ycf4; petG_trnW; petN_psbM; trnC_petN; ycf3_intron2*	5
19	*atpI_rps2; cemA_petA; ndhE_ndhG; rpl32_trnL; rpl33_rps18; rpl36_rps8; trnD_trnY*	7
20	*psbI_trnS; rps8_rpl14; trnR_atpA*	3
21	*rpl16_rps3; trnfM_rps14; trnM_atpE*	3

2.5. IR Expansion and Contraction in Lilium

Based on the topology of *Lilium* species using Bayesian inference with all loci and partition models, the movements of IR boundaries (LSC-IRb, IRb-SSC, and SSC-IRa) were investigated (Figure 5). The LSC-IRb boundaries of group A were identical to each other. In contrast to group A, IR expansions and contractions were found in group B, excluding *L. distichum*, *L. washingtonianum*, *L. pardalinum*, and *L. superbum*. Two IR-SSC boundaries were more diverse than LSC-IRb, although most IR expansions and contractions occurred in the SSC-IRa boundary. Overall, IR boundaries were highly conserved within clades, except clades IV and VI.

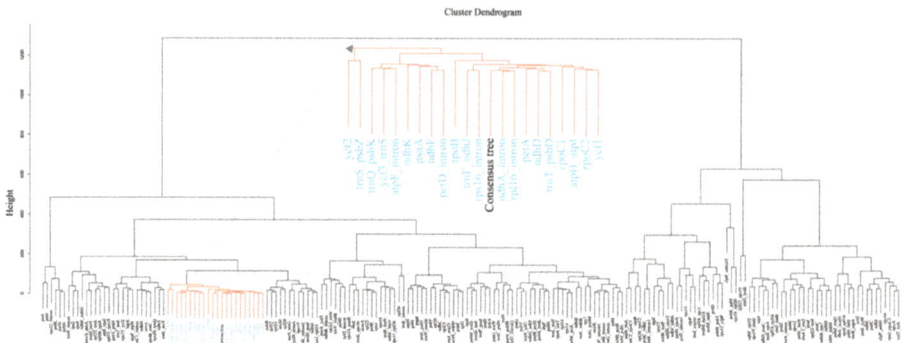

Figure 4. Hierarchical cluster analysis for 190 trees, including 189 gene trees and a consensus tree. The horizontal dashed line refers to the cut-off value for cluster trees.

2.6. Insertions/Deletions in Lilium

In total, eight insertions/deletions (indels) longer than 50 bp occurred in over five species that were found throughout the whole plastome sequences. Five of them had distinguishable features between groups A and B in the *Lilium* phylogeny (Figure S5A), but others did not correspond to the phylogenetic relationships (Figure S5B).

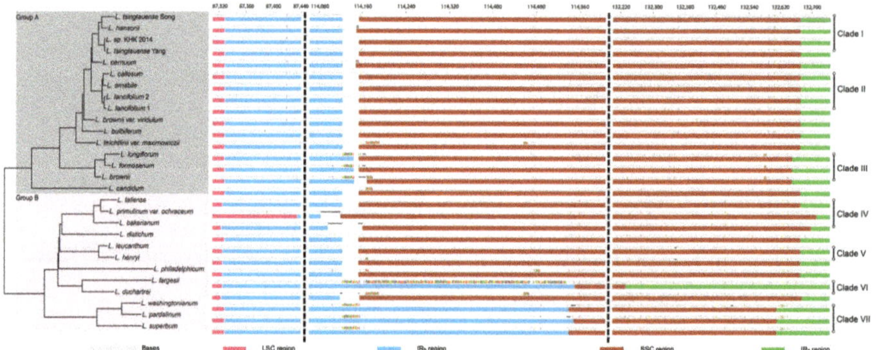

Figure 5. Inverted repeat (IR) expansion and contraction in *Lilium*. Pale grey refers to bases and colors on the bases represent disagreements with the consensus sequence. The red, blue, brown, and green blocks below bases stand for the large single copy (LSC) region, IRb region, small single copy (SSC) region, and IRa region, respectively. The tree on the left is constructed using Bayesian inference with all loci and partition models.

3. Discussion

3.1. Evolution of Lilium Plastomes in Liliaceae

The plastome sequences of *Lilium*, including the three newly determined species in this study, are highly conserved in terms of gene content and order and genomic structure. Although the nucleotide diversities of the LSC and SSC regions were higher than that of the IR region, this is a common phenomenon across the angiosperms [25]. However, the nucleotide diversities of *Lilium* and *Fritillaria* were higher in the LSC and SSC regions but lower in the IR region as compared to that of *Paris* belonging to Melanthiaceae of Liliales [44]. Consequently, the IR regions in Liliaceae appear to be more stable by purifying selection, and the LSC and SSC regions were assumed to have been under relaxed purifying selective pressure compared to those of Melanthiaceae during their evolutionary process.

In Liliaceae, the fluctuating delta nucleotide diversity between *Lilium* and *Fritillaria* also supported the idea that different loci of the plastome have undergone different selective regions in this lineage, except the overestimated nucleotide diversities owing to deletions and small inversions. These findings imply that the mutational dynamics with respect to plastome loci between *Lilium* and *Fritillaria* have been processed differently even though they are closely related taxa.

3.2. Verification of Comber's Sectional Classification

In this study, we reconstructed the phylogeny of 28 *Lilium* species using complete plastomes with four different datasets, four tools, and two models and compared these trees to determine the accurate species tree. There were slightly different topologies, but the major clades were coincident among the trees and were strongly supported by various branch support values (Figure S1). Before the discussion of the phylogeny of *Lilium*, we must evaluate the identification of plastome sequences of *L. distichum* (NC_029937) and *L.* sp. KHK_2014 (NC_027679), which were published by the same research group, because of their unreliable phylogenetic position in the genus. Du et al. [4] first raised a question regarding the phylogenetic position of *L. distichum* (NC_029937). The *rbcL* and *matK* sequences of *L. distichum* (NC_029937) were identical to those of *L. speciosum* (*rbcL*: AB034922.1; *matK*: AB030853, AB049526). On the contrary, *atpB*, *rbcL*, and *ndhF* of *L.* sp. KHK_2014 (NC_027679) were identical to those of *L. distichum* (*atpB*: KC796843, JX903928, KM085888; *rbcL*: JX903238, JN786059, JN417422, KP711933; *ndhF*: JX903509, KM085762). Consequently, it may be necessary to exclude these two plastomes or to consider *L.* sp. KHK_2014 as *L. distichum* to avoid improper conclusions for the phylogenomics of *Lilium*.

All phylogenetic trees using different datasets indicated that *Lilium* species could be divided into two major groups. These two groups were also distinguished by the mutational dynamics of the IR expansion/contraction and large indels (Figure 5 and Figure S5A).

3.2.1. The Phylogenetic Position of Martagon

Martagon consists of five species that are primarily distributed in northeastern Asia and Russia, except *L. martagon*, which ranges widely from central Europe to eastern Siberia. This section had been considered an early-diverging lineage in *Lilium* based on the morphological characteristics: hypogeal and delayed germination, whorled leaves, jointed scales, and heavy seeds [5,45]. In contrast to the morphological analyses, molecular phylogenetic analyses using ITS sequences showed that this section is a more recently derived lineage and is sister to some of the *Sinomartagon* + *Leucolirion* [10,11,13]. Based on these subgeneric relationships, Gao et al. [41] suggested that the ancestor of *Martagon*, *Sinomartagon*, and *Leucolirion* 6b had a distribution within the Hengduan Mountains before *Martagon* separated from *Sinomartagon* + *Leucolirion* 6b approximately 8.8 million years ago. However, based on the plastome sequences, *Martagon* is not a sister to *Sinomartagon* + *Leucolirion* 6b but forms a clade within *Sinomartagon* I (Figure 5). Additionally, the plastome structures of *Lilium* provide further support. The junctions between SCs and IRs of *Martagon* are identical to those of *Sinomartagon* I, whereas they differ from those of *Sinomartagon* 5c and *Leucolirion* 6b (Figure 5). These results imply that the ancestor of *Martagon* diverged from *Sinomartagon* species, i.e., the ancestor of *L. cernuum* in this study, and the divergence time was more recent than the expectation of Gao et al. [41]. In addition, four species within *Sinomartagon* I and three species within *Martagon* are commonly distributed in eastern Asia. As a result, the hypothesis that the origin of *Martagon* is the Hengduan Mountains [41] is controvertible.

3.2.2. The Polyphyly of Pseudolirium

Pseudolirium consists of all American lilies, including *L. philadelphicum*, which is a lectotype of the section [5]. Interestingly, when the data matrix involved *L. philadelphicum*, the phylogeny using ITS [10,11,13] showed the section was monophyletic, but using *matK* [46] revealed polyphyly for the section. Unfortunately, the phylogenetic position of the section using both markers was not resolved or was supported weekly. On the other hand, Kim et al. [47] suggested that *L. philadelphicum* seems to be distinguishable from other species in the section based on phylogenomics using plastome sequences. This relationship is well supported by the 28 species trees constructed in this study. In terms of DNA sequence mutations, a two base deletion in ITS was found specifically in *Pseudolirium* species, except for *L. philadelphicum* [10], and there was an obvious distinction between the IR-SC junctions of *L. philadelphicum* and other *Pseudolirium* species (Figure 5). Morphologically, subsection 2d, consisting of *L. philadelphicum* and *L. catesbaei* in *Pseudolirium*, is distinguished from other species by erect flowers and highly clawed perianth parts [3]. Consequently, a new circumscription of *Pseudolirium* should be considered to reflect the recent phylogenetic results.

3.2.3. The Polyphyly of Leucolirion

Leucolirion consists of eight species with scattered and sessile leaves and trumpet flowers [1]. The section is subdivided into two subsections based on bulb color: dark purple or brown for 6a and white for 6b [1]. In this study, the two subsections were distantly separated with strong support and this result was congruent with the previous phylogenetic studies [13,46]. In addition, *L. henryi* of *Sinomartagon* 5a and *L. brownii* of *Archelirion* formed a robust clade with *Leucolirion* 6a and 6b, respectively (Figure S1, Figure 5). Based on the phylogenetic and cytological studies, Du et al. [13] suggested that *L. henryi* and *L. brownii* should be classified into *Leucolirion* 6a and 6b, respectively. Consequently, our results provide further support for the modification of *Leucolirion* according to Du et al. [13].

3.2.4. The Position of Liriotypus in the Genus *Lilium*

Liriotypus comprises 20 species, including all European, Turkish, and Caucasian species, with the exception of *Lilium martagon* [5,48]. Among them, *L. bulbiferum*, having upright flowers, has been distinguished from the rest of the species within the section and forms a clade with the *Sinomartagon* species, including *L. dauricum* of *Daurolirion* based on the molecular phylogenetic analyses [13,48]. In this study, *L. bulbiferum* was placed far away from *L. candidum*, which is a lectotype of the section [5], agreeing with the results of previous studies. However, it forms a clade with *Martagon* + *Sinomartagon* I with strong support, although there are sampling gaps that prevent a concrete conclusion (Figure S1). Therefore, increased taxa sampling, particularly the members of *Daurolirion*, will help to resolve the discordance of the phylogenetic position of *L. bulbiferum* between this study and previous studies and offer the correct phylogenetic position for this species.

On the other hand, the phylogenetic position of *Liriotypus*, except *L. bulbiferum*, was irregular within the genus, although they formed a clade with the *Sinomartagon* 5c species-*Nomocharis* clade in previous studies [11,13]. On the contrary, *L. candidum* was an early-diverging taxon in group A, without alternative relationships with the rest of group A species in this study (Figure 3, Figure S1). Consequently, the newly suggested phylogenetic position of *Liriotypus* in this study is incongruent with that of previous phylogenetic studies using a few molecular markers. One possible explanation for this discordance is that the position of *L. candidum* does not represent *Liriotypus* in spite of its taxonomic importance by lectotype in the subsection. This species differs from other *Liriotypus* species based on rosette basal leaves and widely trumpet-shaped flowers [48,49], and the geographic circumscription of the sections in *Lilium* by Comber [5] collided with the molecular phylogeny of this study, i.e., *Pseudolirium* and *Liriotypus*. Another scenario is that the origin of *Liriotypus* came from early-diverging *Sinomartagon* (see additional details in the next section) and directly moved to Europe via the Caucasus during evolution.

In spite of low taxon sampling in *Liriotypus*, our result strongly supports the previous conclusions in which *L. bulbiferum* is separated from the rest of the species [48], and it suggests a new phylogenetic position of the section within the robust phylogenetic trees. However, because there still remains uncertainty as to whether the position of *L. candidum* belongs to the *Liriotypus* based on its unique morphological characteristics in the section, more sampling in the section will be needed to solve its position.

3.2.5. Biogeographic History of the Genus *Lilium*

The geographical origin of the tribe Lilieae has been speculated to be the Himalayas + Hengduan Mountains and multiple intercontinental dispersal events for Lilieae were suggested, as well as *Lilium* [41,50]. In this study, we also found long distant dispersals and simpler than the previous speculation. Gao et al. [41] suggested a long-distance dispersal model of *Lilium* based on ITS and *matK* phylogenies, in which there were movements among four regions: Hengduan Mountains, eastern Asia, Europe, including Caucasus, and Northern America. However, from the results in this study, it was suggested that there were two main long dispersals in eastern Asia—Europe in group A and Hengduan Mountains—North America in group B (Figure 3).

To explain these distributions, it was supposed that *Lilium* comprised "*Sinomartagon* and its derivatives." *Sinomartagon* comprises approximately 30 species [11] with epigeal and immediate germination (except *L. henryi*), scattered leaves, an entire bulb, and Turk's cap flowers [5], and it has a distribution from the Hengduan Mountains to eastern Asia. In molecular phylogenetic trees, including those in this study, this section has polyphyly regardless of the types of marker or taxon samplings [12,13,41,46]. Therefore, if the manner for distant dispersals was paved during glacial periods or pollinators traveled a great distance at the beginning of the *Lilium* diversification, certain populations of different species in the *Sinomartagon* could have moved together. Therefore, the adaptations to new circumstances may have led to new populations that morphologically converged. Consequently, populations within new circumstances had similar morphological characteristics and belonged to the same section by morphological classification. This may cause the discordance between classifications based on morphological characteristics and molecular phylogeny. Regarding the basal lineages of group A and B, the "*Sinomartagon* and its derivatives" hypothesis is unclear because of our low taxon sampling or extinction of ancient *Sinomartagon* species. This hypothesis may further confuse the evolutionary history of the genus because (1) nobody has placed *Sinomartagon* as a basal lineage in the *Lilium* based on the morphological characteristics, and (2) most phylogenies of *Lilium* constructed using ITS are not consistent with the present results. In addition, inheritance from single parents, such as a plastome, makes it difficult to detect hybridization events. However, the phylogenetic tree generated in this study was the first robust phylogenetic tree with more than 20 samples. There is no doubt that the most important action for the discussion of phylogenetic relationships or divergence times of certain lineage is the construction of accurate species trees without polytomy and poor support. Therefore, we cannot rule out this "*Sinomartagon* and its derivatives" hypothesis based on the well-resolved phylogenetic tree.

3.3. Molecular Markers for the Phylogeny of Lilium and its Relatives

ITS has been used as a valuable molecular marker for the phylogeny of *Lilium* [10,11,13,41] but the phylogenetic relationships among sections or species were incongruent or were weakly supported. To overcome this problem, we constructed the phylogeny of *Lilium* using whole plastome sequences in this study. However, the production and manipulation of NGS data also require significant time and cost, as well as a higher-level technique than the Sanger sequencing method.

To evaluate which gene trees were similar to the species tree in terms of topology, while reflecting the generic relationship in Liliaceae, 189 gene trees were constructed, and only 20 were found to be consistent with the results presented in previous phylogenetic studies [51,52]. Among them, the *ycf1*, *trnF-ndhJ*, and *trnT-psbD* regions were coincident with highly variable regions of the *Lilium* plastome, as suggested by Du et al. [4].

This could serve as an alternative choice of NGS-based phylogenomics in *Lilium* when we use insufficient conditioned samples, such as very low concentration, fragmentation, or extraction from very old specimens that are difficult for use in the preparation of an NGS library.

4. Materials and Methods

4.1. Plant Materials and DNA Extraction

The bulbs of *L. leichtlinii* var. *maximowiczii* (Wooriseed, Korea) and the seeds of *L. formosanum* (Wageningen University, Netherlands) and *L. candidum* (Royal horticultural society Lily group, UK) were germinated on media at Kyungpook National University of Korea. Genomic DNA was extracted from young fresh leaves using a DNeasy plant mini kit (Qiagen, Valencia, CA, USA), following the manufacturer's protocol.

4.2. Sequencing, Assembly, and Annotation

Genomic DNA was sequenced using the HiSeq 2500 instrument (Illumina, San Diego, CA, USA). The following assembling procedures were implemented using Geneious 10.2.5 [53]. Both ends of raw reads were trimmed with more than a 1% chance of an error per base. Reads exceeding 50 bp in length were extracted and used as raw reads after this step. Raw reads were mapped to the plastome sequence of *Lilium pardalinum* [47] with medium-low sensitivity. Reads were aligned to the reference then *de novo* assembled with zero mismatches and gaps among the reads to generate contigs. Raw reads were realigned to the contigs with zero mismatches and gaps among the reads for up to 100 iterations. The generated contigs were concatenated into a circular form using *de novo* assembled circularizing contigs with matching ends. Finally, the raw reads were mapped to the complete plastome sequence with zero mismatches and gaps among the reads to verify the coverage depths through the genome, because of the fact that many plastome-like sequences distributed in the mitochondrial genome and nuclear genome have relatively low coverage depths compared to that of the plastome.

All of the genes in the three plastome sequences were annotated and compared with those of *L. pardalinum* using Geneious annotation with 90% similarity, then re-checked separately using BLASTP [54] and tRNAscan-SE [55].

4.3. Sequence Diversity and GC Content Analyses

Twenty *Fritillaria* and 28 *Lilium* plastome sequences were aligned by MAFFT [56], then the AT-rich regions were realigned by MUSCLE [57] to increase the alignment accuracy at these regions. The alignment sequences were loaded in R ver. 3.5.1 [58], and the nucleotide diversities of the two genera were analyzed using sliding window analysis (window size = 600 bp, step size = 200 bp) by deleting the sites including at least one missing data point for all sequences. In addition, the GC content was also calculated to compare the relationship between nucleotide diversity and GC content.

4.4. Phylogenetic Analysis

Fifty-five plastome sequences from the genera *Lilium*, *Fritillaria*, *Cardiocrinum*, *Amana*, and *Erythronium* were downloaded from GenBank to construct a phylogeny for *Lilium*. We extracted 78 coding sequences (CDSs), 90 intergenic spaces (IGSs) longer than 100 bp, and 21 intron regions (two regions from *clpP*, *trnK*, and *ycf3*) from each plastome (Table S2).

All of the extracted sequences were aligned using MAFFT [55] according to the region (Table S3) and merged into four datasets as follows: All_loci (including CDSs, IGSs, and introns), CDS_loci, IGS_loci, and Intron_loci. The models for each partition for the four datasets were estimated using PartitionFinder 2 [59], ModelFinder [60], or Jmodeltest 2 [61] for different phylogenetic analyses. The phylogenetic trees were constructed using the RAxML Black Box with 1000 bootstraps [62] in the CIPRES gateway [63], MrBayes with ngen = 10,000,000, samplefreq = 1000, and burninfrac = 0.25 [64], or IQ-TREE with 1000 bootstraps [65]. In total, 189 gene trees were constructed using IQ-TREE with ModelFinder, and these were used to estimate the species trees by ASTRAL [66–68]. All of the details for phylogenies are summarized in Table S2. A consensus tree of 28 phylogenies was constructed by majority rule.

4.5. Comparison of Gene Trees

The generic relationships of each gene tree were compared to those presented in previous phylogenetic studies of Liliaceae [51,52]. Clusters of similar gene trees were identified using the Kendall Colijn metric [69] in treespace [43] with five principal components and a cut-off distance of 100. Phylogeny using selected markers was constructed by IQ-TREE, with 1000 bootstraps for comparison to the consensus tree constructed by 28 phylogenies.

4.6. R Packages for Manipulation of Phylogenies

APE [70], Biostrings [71], dplyr [72], ggplot2 [73], ggtree [74,75], gridExtra [76], pegas [77], phytools [78], tidytree [79], and treespace [43] were used in this study.

Supplementary Materials: The following are available online at http://www.mdpi.com/2223-7747/8/12/547/s1, Figure S1: A clade of *Lilium* species extracted from 28 species trees, Figure S2: Gene trees constructed using 78 coding regions. Coloured line refers to each genus, Figure S3: Gene trees constructed using 90 intergenic spacers. Coloured line refers to each genus, Figure S4: Gene trees were constructed using 21 introns. Coloured line refers to each genus, Figure S5: Large indertions/deletions occurring at least five species with longer than 50 bp. Red line distinguishes between two groups, Table S1: Summary of three plastome sequences, Table S2: List of estimated species trees according to different options, Table S3: 189 loci used for phylogenetic analyses in this paper.

Author Contributions: Performed research design, writing of the manuscript, lab work, and data analysis, H.T.K.; collected samples and edited the manuscript, K.-B.L.; contributed to the writing and editing of the manuscript, J.S.K.

Funding: This research was supported by the Basic Science Research Program through the National Research Foundation of Korea (NRF) funded by the Ministry of Education (Project No. NRF-2017R1D1A1A02018573/ 2016R1D1A1B04932913).

Acknowledgments: We thank Ji-Hyang Park for growing the plants from seeds and bulbs.

Conflicts of Interest: The authors declare that the research was conducted in the absence of any commercial or financial relationships that could be construed as a potential conflict of interest.

References

1. Haw, S.G.; Liang, S.-Y. *The Lilies of China: The Genera Lilium, Cardiocrinum, Nomocharis and Notholirion*; Timber Press: Portland, OR, USA, 1986.

2. McRae, E.A.; Austin-Mcrae, E.; MacRae, E. *Lilies: A Guide for Growers and Collectors*; Timber Press: Portland, OR, USA, 1998; Volume 105.

3. Lim, K.-B.; van Tuyl, J.M.L. *Flower Breeding and Genetics*; Springer: Berlin, Germany, 2007; pp. 517–537.

4. Du, Y.-P.; Bi, Y.; Yang, F.-P.; Zhang, M.-F.; Chen, X.-Q.; Xue, J.; Zhang, X.-H. Complete chloroplast genome sequences of *Lilium*: Insights into evolutionary dynamics and phylogenetic analyses. *Sci. Rep.* **2017**, *7*, 5751. [CrossRef] [PubMed]

5. Comber, H.F. *A New Classification of the Genus Lilium*; Royal Horticultural Society: London, UK, 1949; Volume 13, pp. 85–105.

6. Endlicher, S. Genera plantarum. Vindobonae. Apud Fr. Beck Universitatis Bibliopolam 1836. Available online: https://bibdigital.rjb.csic.es/records/item/10951-redirection (accessed on 25 November 2019).

7. Reichenbach, H. Flora Germanica Excursoria.–Leipzig. 1830. Available online: https://www.biodiversitylibrary.org/item/7359#page/3/mode/1up (accessed on 25 November 2019).

8. Baker, J. The Gardeners' Chronicle and and agricultural gazette. 1871. Available online: https://www.biodiversitylibrary.org/page/16448237#page/1163/mode/1up (accessed on 27 November 2019).

9. De Jong, P. Some notes on the evolution of lilies. *Lily Yearb. N. Am. Lily Soc.* **1974**, *27*, 23–28.

10. Nishikawa, T.; Okazaki, K.; Uchino, T.; Arakawa, K.; Nagamine, T. A molecular phylogeny of *Lilium* in the internal transcribed spacer region of nuclear ribosomal DNA. *J. Mol. Evol.* **1999**, *49*, 238–249. [CrossRef] [PubMed]

11. Nishikawa, T.; Okazaki, K.; Arakawa, K.; Nagamine, T. Phylogenetic analysis of section Sinomartagon in genus *Lilium* using sequences of the internal transcribed spacer region in nuclear ribosomal DNA. *Breed. Sci.* **2001**, *51*, 39–46. [CrossRef]

12. Lee, C.S.; Kim, S.-C.; Yeau, S.H.; Lee, N.S. Major lineages of the genus *Lilium* (Liliaceae) based on nrDNA ITS sequences, with special emphasis on the Korean species. *J. Plant Biol.* **2011**, *54*, 159–171. [CrossRef]

13. Du, Y.P.; He, H.B.; Wang, Z.X.; Li, S.; Wei, C.; Yuan, X.N.; Cui, Q.; Jia, G.X. Molecular phylogeny and genetic variation in the genus *Lilium* native to China based on the internal transcribed spacer sequences of nuclear ribosomal DNA. *J. Plant Res.* **2014**, *127*, 249–263. [CrossRef]

14. Du Puy, D.; Cribb, P. *The Genus Cymbidium*; Royal Botanic Gardens: Sydney, Australia, 2007.

15. Sharma, S.K.; Dkhar, J.; Kumaria, S.; Tandon, P.; Rao, S.R. Assessment of phylogenetic inter-relationships in the genus *Cymbidium* (Orchidaceae) based on internal transcribed spacer region of rDNA. *Gene* **2012**, *495*, 10–15. [CrossRef]

16. van den Berg, C.; Ryan, A.; Cribb, P.J.; Chase, M.W. Molecular phylogenetics of *Cymbidium* (Orchidaceae: Maxillarieae): Sequence data from internal transcribed spacers (ITS) of nuclear ribosomal DNA and plastid matK. *Lindleyana* **2002**, *17*, 102–111.

17. Yukawa, T.; Miyoshi, K.; Yokoyama, J. Molecular phylogeny and character evolution of *Cymbidium* (Orchidaceae). *Bull. Natl. Sci. Mus. Tokyo B* **2002**, *28*, 129–139.

18. Zwickl, D.J.; Hillis, D.M. Increased taxon sampling greatly reduces phylogenetic error. *Syst. Biol.* **2002**, *51*, 588–598. [CrossRef]

19. Graybeal, A. Is it better to add taxa or characters to a difficult phylogenetic problem? *Syst. Biol.* **1998**, *47*, 9–17. [CrossRef] [PubMed]

20. Heath, T.A.; Hedtke, S.M.; Hillis, D.M. Taxon sampling and the accuracy of phylogenetic analyses. *J. Syst. Evol.* **2008**, *46*, 239–257. [CrossRef]

21. Rokas, A.; Carroll, S.B. More genes or more taxa? The relative contribution of gene number and taxon number to phylogenetic accuracy. *Mol. Biol. Evol.* **2005**, *22*, 1337–1344. [CrossRef]

22. Rosenberg, M.S.; Kumar, S. Incomplete taxon sampling is not a problem for phylogenetic inference. *Proc. Natl. Acad. Sci. USA* **2001**, *98*, 10751–10756. [CrossRef] [PubMed]

23. Mower, J.P.; Sloan, D.B.; Alverson, A.J. Plant mitochondrial genome diversity: The genomics revolution. In *Plant Genome Diversity Volume 1*; Springer: Berlin, Germany, 2012; pp. 123–144.

24. Gregory, T.R.; Nicol, J.A.; Tamm, H.; Kullman, B.; Kullman, K.; Leitch, I.J.; Murray, B.G.; Kapraun, D.F.; Greilhuber, J.; Bennett, M.D. Eukaryotic genome size databases. *Nucleic Acids Res.* **2006**, *35*, D332–D338. [CrossRef] [PubMed]

25. Ruhlman, T.A.; Jansen, R.K. The plastid genomes of flowering plants. *Methods Mol. Biol.* **2014**, *1132*, 3–38. [CrossRef]

26. Weng, M.L.; Blazier, J.C.; Govindu, M.; Jansen, R.K. Reconstruction of the ancestral plastid genome in Geraniaceae reveals a correlation between genome rearrangements, repeats, and nucleotide substitution rates. *Mol. Biol. Evol.* **2014**, *31*, 645–659. [CrossRef]

27. Guisinger, M.M.; Kuehl, J.V.; Boore, J.L.; Jansen, R.K. Extreme reconfiguration of plastid genomes in the angiosperm family Geraniaceae: Rearrangements, repeats, and codon usage. *Mol. Biol. Evol.* **2011**, *28*, 583–600. [CrossRef]

28. Westwood, J.H.; Yoder, J.I.; Timko, M.P.; dePamphilis, C.W. The evolution of parasitism in plants. *Trends Plant Sci.* **2010**, *15*, 227–235. [CrossRef]

29. Shaw, J.; Lickey, E.B.; Schilling, E.E.; Small, R.L. Comparison of whole chloroplast genome sequences to choose noncoding regions for phylogenetic studies in angiosperms: The tortoise and the hare III. *Am. J. Bot.* **2007**, *94*, 275–288. [CrossRef]

30. Shaw, J.; Lickey, E.B.; Beck, J.T.; Farmer, S.B.; Liu, W.; Miller, J.; Siripun, K.C.; Winder, C.T.; Schilling, E.E.; Small, R.L. The tortoise and the hare II: Relative utility of 21 noncoding chloroplast DNA sequences for phylogenetic analysis. *Am. J. Bot.* **2005**, *92*, 142–166. [CrossRef] [PubMed]

31. Yang, J.B.; Tang, M.; Li, H.T.; Zhang, Z.R.; Li, D.Z. Complete chloroplast genome of the genus *Cymbidium*: Lights into the species identification, phylogenetic implications and population genetic analyses. *BMC Evol. Biol.* **2013**, *13*, 84. [CrossRef]

32. Zhang, Y.J.; Ma, P.F.; Li, D.Z. High-throughput sequencing of six bamboo chloroplast genomes: Phylogenetic implications for temperate woody bamboos (Poaceae: Bambusoideae). *PLoS ONE* **2011**, *6*, e20596. [CrossRef] [PubMed]

33. Wysocki, W.P.; Clark, L.G.; Attigala, L.; Ruiz-Sanchez, E.; Duvall, M.R. Evolution of the bamboos (Bambusoideae; Poaceae): A full plastome phylogenomic analysis. *BMC Evol. Biol.* **2015**, *15*, 50. [CrossRef] [PubMed]

34. Dong, W.; Liu, J.; Yu, J.; Wang, L.; Zhou, S. Highly variable chloroplast markers for evaluating plant phylogeny at low taxonomic levels and for DNA barcoding. *PLoS ONE* **2012**, *7*, e35071. [CrossRef] [PubMed]

35. Sarkinen, T.; George, M. Predicting plastid marker variation: Can complete plastid genomes from closely related species help? *PLoS ONE* **2013**, *8*, e82266. [CrossRef] [PubMed]

36. Ku, C.; Hu, J.M.; Kuo, C.H. Complete plastid genome sequence of the basal asterid Ardisia polysticta Miq. and comparative analyses of asterid plastid genomes. *PLoS ONE* **2013**, *8*, e62548. [CrossRef] [PubMed]

37. Ku, C.; Chung, W.C.; Chen, L.L.; Kuo, C.H. The complete plastid genome sequence of madagascar periwinkle *Catharanthus roseus* (L.) G. Don: Plastid genome evolution, molecular marker identification, and phylogenetic implications in asterids. *PLoS ONE* **2013**, *8*, e68518. [CrossRef]

38. Kim, H.T.; Kim, J.S.; Lee, Y.M.; Mun, J.-H.; Kim, J.-H. Molecular markers for phylogenetic applications derived from comparative plastome analysis of Prunus. *J. Syst. Evol.* **2019**, *57*, 15–22. [CrossRef]

39. Kim, J.-H.; Lee, S.-I.; Kim, B.-R.; Choi, I.-Y.; Ryser, P.; Kim, N.-S. Chloroplast genomes of *Lilium lancifolium*, *L. amabile*, *L. callosum*, and *L. philadelphicum*: Molecular characterization and their use in phylogenetic analysis in the genus *Lilium* and other allied genera in the order Liliales. *PLoS ONE* **2017**, *12*, e0186788. [CrossRef]

40. Maddison, W.P. Gene trees in species trees. *Syst. Biol.* **1997**, *46*, 523–536. [CrossRef]

41. Gao, Y.D.; Harris, A.J.; Zhou, S.D.; He, X.J. Evolutionary events in *Lilium* (including Nomocharis, Liliaceae) are temporally correlated with orogenies of the Q-T plateau and the Hengduan Mountains. *Mol. Phylogenet. Evol.* **2013**, *68*, 443–460. [CrossRef] [PubMed]

42. Yu, G. A tidy tool for phylogenetic tree data manipulation [R package tidytree version 0.1. 9]. *Compr. R Arch. Netw.* 2019. Available online: https://rdrr.io/cran/tidytree/ (accessed on 25 November 2019).

43. Jombart, T.; Kendall, M.; Almagro-Garcia, J.; Colijn, C. Treespace: Statistical exploration of landscapes of phylogenetic trees. *Mol. Ecol. Resour.* **2017**, *17*, 1385–1392. [CrossRef] [PubMed]

44. Song, Y.; Wang, S.; Ding, Y.; Xu, J.; Li, M.F.; Zhu, S.; Chen, N. Chloroplast genomic resource of Paris for species discrimination. *Sci. Rep.* **2017**, *7*, 3427. [CrossRef] [PubMed]

45. Lighty, R. The lilies of Korea. *Lily Yearb. RHS* **1969**, *31*, 31–39.

46. Hayashi, K.; Kawano, S. Molecular systematics of *Lilium* and allied genera (Liliaceae): Phylogenetic relationships among *Lilium* and related genera based on the rbcL and matK gene sequence data. *Plant Species Biol.* **2000**, *15*, 73–93. [CrossRef]

47. Kim, H.T.; Zale, P.J.; Lim, K.-B. Complete plastome sequence of *Lilium pardalinum* Kellogg (Liliaceae). *Mitochondrial DNA Part B* **2018**, *3*, 478–479. [CrossRef]

48. İkinci, N.; Oberprieler, C.; Güner, A. On the origin of European lilies: Phylogenetic analysis of *Lilium* section Liriotypus (Liliaceae) using sequences of the nuclear ribosomal transcribed spacers. *Willdenowia* **2006**, *36*, 647–656. [CrossRef]

49. Muratović, E.; Hidalgo, O.; Garnatje, T.; Siljak-Yakovlev, S. Molecular phylogeny and genome size in European lilies (Genus *Lilium*, Liliaceae). *Adv. Sci. Lett.* **2010**, *3*, 180–189. [CrossRef]

50. Patterson, T.B.; Givnish, T.J. Phylogeny, concerted convergence, and phylogenetic niche conservatism in the core Liliales: Insights from rbcL and ndhF sequence data. *Evolution* **2002**, *56*, 233–252. [CrossRef]

51. Rønsted, N.; Law, S.; Thornton, H.; Fay, M.F.; Chase, M.W. Molecular phylogenetic evidence for the monophyly of *Fritillaria* and *Lilium* (Liliaceae; Liliales) and the infrageneric classification of *Fritillaria*. *Mol. Phylogenet. Evol.* **2005**, *35*, 509–527. [CrossRef] [PubMed]

52. Fay, M.F.; Chase, M.W.; Rønsted, N.; Devey, D.S.; Pillon, Y.; Pires, J.C.; Peterson, G.; Seberg, O.; Davis, J.I. Phylogenetics of Lilliales. *Aliso J. Syst. Evol. Bot.* **2006**, *22*, 559–565. [CrossRef]

53. Kearse, M.; Moir, R.; Wilson, A.; Stones-Havas, S.; Cheung, M.; Sturrock, S.; Buxton, S.; Cooper, A.; Markowitz, S.; Duran, C.; et al. Geneious Basic: An integrated and extendable desktop software platform for the organization and analysis of sequence data. *Bioinformatics* **2012**, *28*, 1647–1649. [CrossRef] [PubMed]

54. Altschul, S.F.; Gish, W.; Miller, W.; Myers, E.W.; Lipman, D.J. Basic local alignment search tool. *J. Mol. Biol.* **1990**, *215*, 403–410. [CrossRef]

55. Lowe, T.M.; Eddy, S.R. tRNAscan-SE: A program for improved detection of transfer RNA genes in genomic sequence. *Nucleic Acids Res.* **1997**, *25*, 955–964. [CrossRef]

56. Katoh, K.; Misawa, K.; Kuma, K.; Miyata, T. MAFFT: A novel method for rapid multiple sequence alignment based on fast Fourier transform. *Nucleic Acids Res.* **2002**, *30*, 3059–3066. [CrossRef]

57. Edgar, R.C. MUSCLE: A multiple sequence alignment method with reduced time and space complexity. *BMC Bioinform.* **2004**, *5*, 113. [CrossRef]

58. Team, R.C. *R: A Language and Environment for Statistical Computing*; R Foundation for Statistical Computing: Vienna, Austria, 2018.

59. Lanfear, R.; Frandsen, P.B.; Wright, A.M.; Senfeld, T.; Calcott, B. PartitionFinder 2: New methods for selecting partitioned models of evolution for molecular and morphological phylogenetic analyses. *Mol. Biol. Evol.* **2016**, *34*, 772–773. [CrossRef]

60. Kalyaanamoorthy, S.; Minh, B.Q.; Wong, T.K.; von Haeseler, A.; Jermiin, L.S. ModelFinder: Fast model selection for accurate phylogenetic estimates. *Nat. Methods* **2017**, *14*, 587. [CrossRef]

61. Darriba, D.; Taboada, G.L.; Doallo, R.; Posada, D. jModelTest 2: More models, new heuristics and parallel computing. *Nat. Methods* **2012**, *9*, 772. [CrossRef]

62. Stamatakis, A. RAxML version 8: A tool for phylogenetic analysis and post-analysis of large phylogenies. *Bioinformatics* **2014**, *30*, 1312–1313. [CrossRef] [PubMed]

63. Miller, M.A.; Pfeiffer, W.; Schwartz, T. Creating the CIPRES Science Gateway for inference of large phylogenetic trees. In Proceedings of the 2010 Gateway Computing Environments Workshop (GCE), New Orleans, LA, USA, 14 November 2010; pp. 1–8.

64. Ronquist, F.; Teslenko, M.; van der Mark, P.; Ayres, D.L.; Darling, A.; Hohna, S.; Larget, B.; Liu, L.; Suchard, M.A.; Huelsenbeck, J.P. MrBayes 3.2: Efficient Bayesian phylogenetic inference and model choice across a large model space. *Syst. Biol.* **2012**, *61*, 539–542. [CrossRef]

65. Nguyen, L.-T.; Schmidt, H.A.; von Haeseler, A.; Minh, B.Q. IQ-TREE: A fast and effective stochastic algorithm for estimating maximum-likelihood phylogenies. *Mol. Biol. Evol.* **2014**, *32*, 268–274. [CrossRef]

66. Zhang, C.; Rabiee, M.; Sayyari, E.; Mirarab, S. ASTRAL-III: Polynomial time species tree reconstruction from partially resolved gene trees. *BMC Bioinform.* **2018**, *19*, 153. [CrossRef]

67. Mirarab, S.; Warnow, T. ASTRAL-II: Coalescent-based species tree estimation with many hundreds of taxa and thousands of genes. *Bioinformatics* **2015**, *31*, i44–i52. [CrossRef]

68. Mirarab, S.; Reaz, R.; Bayzid, M.S.; Zimmermann, T.; Swenson, M.S.; Warnow, T. ASTRAL: Genome-scale coalescent-based species tree estimation. *Bioinformatics* **2014**, *30*, i541–i548. [CrossRef]

69. Kendall, M.; Colijn, C. A tree metric using structure and length to capture distinct phylogenetic signals. *arXiv* **2015**, arXiv:preprint/1507.05211.

70. Paradis, E.; Claude, J.; Strimmer, K. APE: Analyses of phylogenetics and evolution in R language. *Bioinformatics* **2004**, *20*, 289–290. [CrossRef]

71. Pagès, H.; Aboyoun, P.; Gentleman, R.; DebRoy, S. Biostrings: Efficient manipulation of biological strings. *R Package Version* **2018**, *2*. Available online: https://rdrr.io/bioc/Biostrings/ (accessed on 25 November 2019).

72. Wickham, H.; Francois, R.; Henry, L.; Müller, K. Dplyr: A grammar of data manipulation. *R Package Version* **2018**, *3*. Available online: https://rdrr.io/cran/dplyr/ (accessed on 25 November 2019).

73. Wickham, H. *Ggplot2: Elegant Graphics for Data Analysis*; Springer: Berlin, Germany, 2016.

74. Yu, G.; Smith, D.K.; Zhu, H.; Guan, Y.; Lam, T.T.Y. Ggtree: An R package for visualization and annotation of phylogenetic trees with their covariates and other associated data. *Methods Ecol. Evol.* **2017**, *8*, 28–36. [CrossRef]

75. Yu, G.; Lam, T.T.-Y.; Zhu, H.; Guan, Y. Two methods for mapping and visualizing associated data on phylogeny using ggtree. *Mol. Biol. Evol.* **2018**, *35*, 3041–3043. [CrossRef]

76. Auguie, B. GridExtra: Functions in Grid graphics. *R Package Version 0.9* **2012**, *1*. Available online: http://CRAN.R-project.org/package=gridExtra (accessed on 25 November 2019).

77. Paradis, E. Pegas: An R package for population genetics with an integrated–modular approach. *Bioinformatics* **2010**, *26*, 419–420. [CrossRef]

78. Revell, L.J. Phytools: An R package for phylogenetic comparative biology (and other things). *Methods Ecol. Evol.* **2012**, *3*, 217–223. [CrossRef]

79. Xinqi, C.; Songyun, L.; Jiemei, X.; Tamura, M. Liliaceae. *Flora China* **2000**, *24*, 73–263.

Article

Organization Features of the Mitochondrial Genome of Sunflower (*Helianthus annuus* L.) with ANN2-Type Male-Sterile Cytoplasm

Maksim S. Makarenko [1,2,*], Alexander V. Usatov [1], Tatiana V. Tatarinova [2,3,4,5], Kirill V. Azarin [1], Maria D. Logacheva [2,6], Vera A. Gavrilova [7], Igor V. Kornienko [1,8] and Renate Horn [9]

[1] Department of Genetics, Southern Federal University, Rostov-on-Don 344006, Russia; usatova@sfedu.ru (A.V.U.); azarinkv@sfedu.ru (K.V.A.); ivkornienko@sfedu.ru (I.V.K.)

[2] The Institute for Information Transmission Problems, Moscow 127051, Russia; ttatarinova@laverne.edu (T.V.T.); m.logacheva@skoltech.ru (M.D.L.)

[3] Department of Biology, University of La Verne, La Verne, CA 91750, USA

[4] Vavilov Institute of General Genetics, Moscow 119333, Russia

[5] School of Fundamental Biology and Biotechnology, Siberian Federal University, Krasnoyarsk 660041, Russia

[6] Skolkovo Institute of Science and Technology, Moscow 121205, Russia

[7] The N.I. Vavilov All-Russian Institute of Plant Genetic Resources, Saint Petersburg 190121, Russia; v.gavrilova@vir.nw.ru

[8] Southern Scientific Center of the Russian Academy of Sciences, Rostov-on-Don 344006, Russia

[9] Institute of Biological Sciences, Plant Genetics, University of Rostock, 18059 Rostock, Germany; renate.horn@uni-rostock.de

* Correspondence: mcmakarenko@yandex.ru

Received: 18 August 2019; Accepted: 19 October 2019; Published: 23 October 2019

Abstract: This study provides insights into the flexibility of the mitochondrial genome in sunflower (*Helianthus annuus* L.) as well as into the causes of ANN2-type cytoplasmic male sterility (CMS). *De novo* assembly of the mitochondrial genome of male-sterile HA89(ANN2) sunflower line was performed using high-throughput sequencing technologies. Analysis of CMS ANN2 mitochondrial DNA sequence revealed the following reorganization events: twelve rearrangements, seven insertions, and nine deletions. Comparisons of coding sequences from the male-sterile line with the male-fertile line identified a deletion of *orf777* and seven new transcriptionally active open reading frames (ORFs): *orf324*, *orf327*, *orf345*, *orf558*, *orf891*, *orf933*, *orf1197*. Three of these ORFs represent chimeric genes involving *atp6* (*orf1197*), *cox2* (*orf558*), and *nad6* (*orf891*). In addition, *orf558*, *orf891*, *orf1197*, as well as *orf933*, encode proteins containing membrane domain(s), making them the most likely candidate genes for CMS development in ANN2. Although the investigated CMS phenotype may be caused by simultaneous action of several candidate genes, we assume that *orf1197* plays a major role in developing male sterility in ANN2. Comparative analysis of mitogenome organization in sunflower lines representing different CMS sources also allowed identification of reorganization hot spots in the mitochondrial genome of sunflower.

Keywords: sunflower; cytoplasmic male sterility (CMS); mitochondrial genome; reorganizations; next generation sequencing (NGS)

---

1. Introduction

Low substitution rate in genes along with considerable variability in size and structure are distinct features of plant mitochondrial genomes (mitogenome) [1,2]. Reorganization events in mitochondrial DNA (mtDNA) are primarily caused by disruption of a fragile equilibrium of intramolecular recombinations, maintained by nuclear-mitochondrial genetic interactions [3,4].

Runaway recombination of mtDNA can lead to changes in gene content and expression patterns of mitochondria [5,6]. The mitogenomes of flowering plants carry genes for rRNAs, tRNA, subunits of the respiratory chain complexes, as well as genes for the ribosomal proteins (*rps* and *rpl*). Maturase-related protein gene (*matR*) and genes responsible for the biogenesis of cytochrome *c* (*ccmB*, *ccmC*, *ccmFC*, and *ccmFN*) are also part of the plant mitochondrial gene set [7]. Alterations in transcription activity of the mitochondrial genes can have profound effects on the functionality of mitochondria and, thus, on different plant traits. Among the phenotypic traits caused by mitochondrial impairments, special attention is devoted to cytoplasmic male sterility (CMS) [8]. CMS is an inability to produce or shed functional pollen, which has been described in more than 140 higher plant species [9]. As a result of mtDNA rearrangements, new open reading frames are created, leading to male sterility [10]. In turn, dominant nuclear-encoded restorer-of-fertility genes (*Rf* genes) can restore normal development of pollen. Hence CMS/Rf systems are important both for studying pollen development in plants and for commercial applications [11,12]. The existence of the CMS phenotype in plants eliminates the need for laborious, manual emasculations for a directional crossing of plants, thus promoting its utilization in hybrid breeding [13].

Comparing mtDNA configurations in sunflower is especially interesting, as more than 70 CMS sources have been described within the Food and Agriculture Organization (F.A.O.) program for sunflower [14], even though modern sunflower hybrid breeding predominantly relies on a single male-sterile cytoplasm, the so-called CMS PET1 [15]. This CMS source originated from an interspecific cross of *H. petiolaris* with *H. annuus* [16]. The molecular characterization of the CMS mechanism helps to introduce new CMS sources into breeding programs. So far, the mitogenomes of only a few CMS sources have been sufficiently characterized to be used in sunflower hybrid production. The CMS phenotype can arise spontaneously in wild populations, while in breeding lines—after wide crosses, interspecific exchange of nuclear and/or cytoplasmic genomes, and mutagenesis [11]. It has been demonstrated that some CMS sources obtained from different inter- or intraspecific crosses showed the same mechanism of male sterility formation as the CMS PET1 type [17]. Even though these CMS sources had different origins, they have the same mitochondrial genome organization indicating that some configurations may be preferentially maintained in sunflowers [17].

Less is known about the spontaneously occurring CMS sources in *Helianthus annuus* L. [14]. ANN2 was derived from wild sunflower population N517 in Texas [18]. In this population, 40% of the plants were male-sterile [19]. However, sunflower cultivars with ANN2 CMS type, developed from the N517 population by maintaining with lines like HA89 or RHA265, showed 100% male-sterile progenies, which indicates a stable mitochondrial DNA configuration and absence of heteroplasmy. ANN2 and other spontaneously occurring CMS sources like ANN1, ANN3, and ANN4 maintained by RHA265 were hardly restored [20]. None of the tested maintainer and restorer lines of CMS PET1 were able to restore pollen production in CMS ANN1 and CMS ANN3. Only 12.5% and 15.8% of all investigated lines showed restorer capacity towards CMS ANN2 and CMS ANN4, respectively, indicating very different CMS mechanisms compared to CMS PET1. Three restorer lines, Rf ANN2-PI 413178, Rf ANN2-P21, and Rf ANN2-RMAX1, carry a restorer-of-fertility gene for ANN2 [21,22]. A suppressor gene S1 overpowering the restorer gene action has been recently described by Liu et al. [23], thus making the CMS-Rf interactions in the ANN2 source even more complicated.

Previous mtDNA investigations of some spontaneously occurring CMS sources were based on Southern blot hybridizations with mitochondrial genes [24]. Hybridizations of the CMS sources ANN1, ANN2, ANN3 with *atp6*, *atp9*, *cob*, *cox1*, *cox2*, *cox3*, *rrn18/rrn5/nad5*, *orfH522*, *orfH708*, or *orfH873* as a probe revealed unique banding patterns for 4 out of the 10 probes [24]. Besides, the analyses showed that CMS ANN4 and ANN5 are very similar to each other and form a distinct group from CMS ANN1, ANN2, and ANN3 [24]. It was also shown that ANN1/ANN2/ANN3 mtDNA-type significantly differs from both the male-fertile sunflower cytoplasm and the CMS PET1 source [24].

We describe the first assembly of the CMS source ANN2, which occurred spontaneously in *Helianthus annuus* L. The current study also provides insights into the flexibility of sunflower mitochondrial genome by comparing different isonuclear male-sterile lines HA89 (ANN2, MAX1, PET1, PET2) and the male-fertile line HA89, allowing identification of hot spots for rearrangements in the sunflower mitochondrial genome. For the CMS mechanism in ANN2, new open reading frames were identified, which were transcriptionally active. The ANN2 CMS source may be interesting not only for oilseed hybrid breeding, but also for horticultural purposes, as it is difficult to restore. It is a highly desirable trait in ornamental sunflowers since pollen production is usually not required nor looked-for, except if the pollen color enhances the contrast with the florets.

2. Results

2.1. Rearrangements in the Mitochondrial Genome of the Male-Sterile Line HA89(ANN2)

We assembled the complete mitochondrial genome of the HA89 sunflower line with ANN2 cytoplasmic sterility type (NCBI accession MN175741.1). The master chromosome of HA89 (ANN2) consists of 306,018 bp (Figure 1), and it is 5071 bp longer than the mitogenome of the male-fertile isonuclear line HA89 (NCBI accession MN171345.1).

The HA89 (ANN2) mitochondrial genome has a wide range of rearrangements as compared to the male-fertile HA89 mitogenome. The summary of whole mitogenome alignment of male-sterile and male-fertile lines is presented in Table 1.

Table 1. Alignment of the mitochondrial genomes of HA89 and HA89(ANN2) lines.

№ of Alignment Region	The Alignment Region Length, bp	Positions in mtDNA of Male-Fertile HA89	Positions in mtDNA of Male-Sterile HA89(ANN2)	Orientation	Similarity %	Localized Genes
1	29,196	1–29,196	1–29,204	Plus/Plus	99	*nad2_ex3,4, trnY, trnN, trnC, ccmC, trnT, atp4, nad4L*
2	557	33,772–34,328	78,343–78,899	Plus/Plus	99	-
3	1245	34,329–35,573	148,163–149,411	Plus/Minus	98	-
4	77,441	36,739–114,179	217,575–295,553	Plus/Plus	95	*atp8, cox3, trnV, rpl5, nad4, trnD, trnK, ccmB, rpl10, trnM, trnG, trnQ, trnH, trnE, nad1_ex1, cox1, nad5_ex3,4, rps11*
5	41,702	114,180–155,882	35,657–77,315	Plus/Minus	99	*atp9, trnM, rps4, rrn26, rrn5, rrn18, rps13, nad1_ex2,3*
6	4150	155,883–160,032	300,892–305,041	Plus/Plus	99	-
7	8584	160,320–168,903	129,358–137,946	Plus/Plus	99	*nad2_ex1,2, nad6 **
8	21,433	168,906–190,275	171,388–192,871	Plus/Minus	98	*nad6 *, trnP, trnF, trnS, trnM, mttB, cob*
9	8158	194,543–202,700	163,232–171,387	Plus/Plus	99	*ccmFC, orf873*
10	24,687	202,701–227,387	192,915–217,574	Plus/Minus	99	*atp1, ccmFN, nad7*
11	41,505	227,396–268,900	87,945–129,446	Plus/Plus	99	*nad1_ex4, rps3, rpl16, trnM, matR, nad3, rps12, nad9, trnW, nad5_ex1,2*
12	6029	269,217–275,245	141,704–147,732	Plus/Minus	99	*atp6 **
13	12,520	275,536–288,055	150,723–163,231	Plus/Plus	99	*cox2*
14	977	299,971–300,947	305,042–306,018	Plus/Plus	99	*trnK*

* genes, which had impaired sequences as a result of rearrangements.

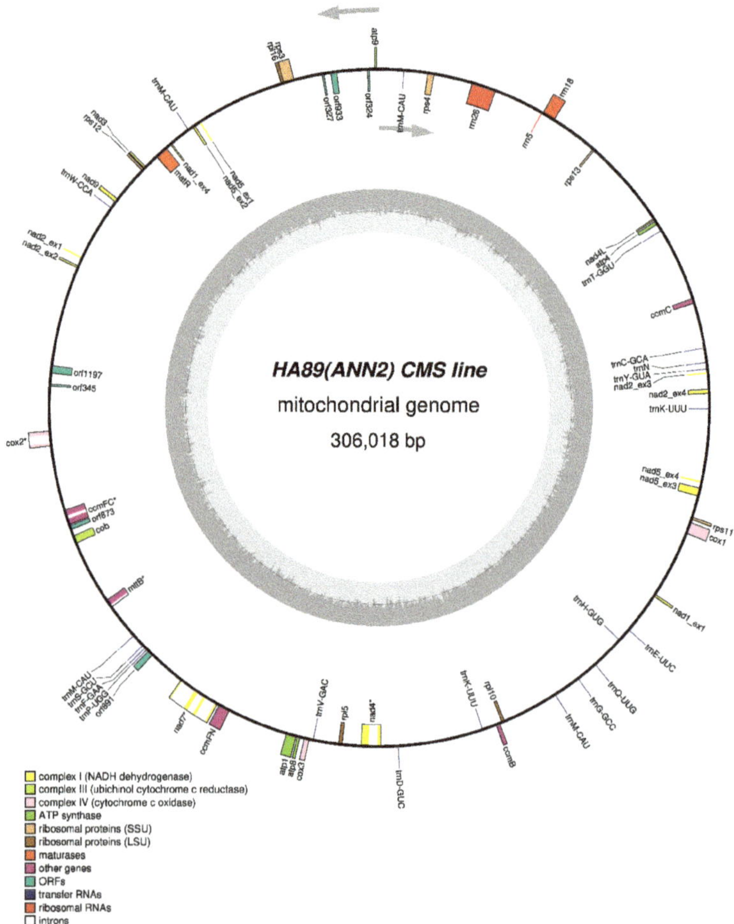

Figure 1. Mitochondrial genome map of HA89(ANN2) cytoplasmic male sterility (CMS) line of sunflower. Intron containing genes are marked by an asterisk (*) symbol. Trans-spliced genes are presented as the compilation of exons (ex).

The mitochondrial genomes of male-sterile ANN2 and male-fertile HA89 share 14 complementary regions, but their localizations and orientation differ. We showed the localization of complementary regions in a scheme with both genomes shown in linear forms in Figure 2. Since regions #1 and #14 in the case of the circular molecule represent the same region, we classified the other twelve regions (#2–#13) as rearrangements.

In most cases, rearrangements only resulted in a reversal of a gene's direction or a change in gene order. However, the 8584 bp (#7) and 21,433 bp (#8) rearrangements influenced the coding sequence of *nad6*, and the 6029 bp rearrangement (#12) impaired *atp6*. The largest part of the *nad6* gene sequence (~88%) is in the rearrangement #8, while the 3′ terminal part of *nad6* lies in the rearrangement #7. As a result of the convergence of #8 and #10 rearrangements in mitochondrial DNA of HA89(ANN2), the new *nad6*-chimeric open reading frame—*orf891*—was created. Analyses of *orf891* transcription pattern gave ambiguous results. Transcripts were detected for both *nad6* (HA89) and *orf891* (HA89(ANN2)) when using primers derived from the 5′ identical sequence of their mRNA (Supplementary Table S1). Nevertheless, using the same forward primer, but different reverse primers

(Supplementary Table S1), complementary to the 3′ sequence of *nad6* and *orf891*, transcription was detected only for *nad6* (the fertile line), but not for *orf891* (CMS line). It is important to note that almost all the rearrangements found in mtDNA of HA89(ANN2) are accompanied by other types of genome reorganizations—deletions and insertions.

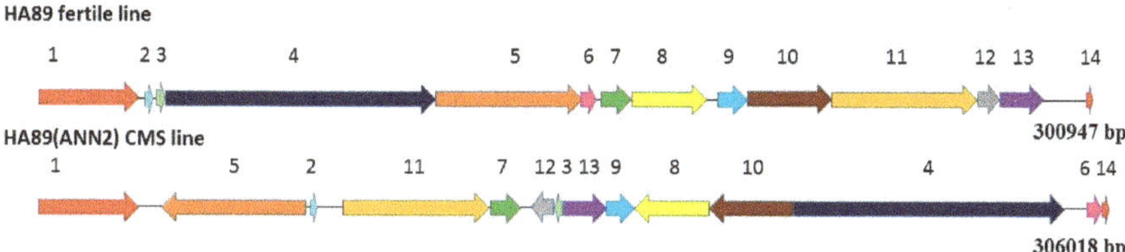

Figure 2. Schematic illustration of homologous regions between mitochondrial genomes of HA89 male-fertile and HA89(ANN2) CMS lines. 1—29,196 bp; 2—557 bp; 3—1245 bp; 4—77,441 bp; 5—41,702 bp; 6—4150 bp; 7—8584 bp; 8—21,433 bp 9—8158 bp; 10—24,687 bp; 11—41,505 bp; 12—6029 bp; 13—12,520 bp; 14—977 bp.

2.2. Deletions and Insertions in the Mitochondrial Genome of the Male-Sterile Line HA89(ANN2)

In comparison to the male-fertile analog, we identified nine longer than 100 bp deletions in the mtDNA of HA89(ANN2), which are shown in Table 2. Most deletions did not affect the protein-coding sequences, except for two deletions of 316 bp and 1165 bp. The 1165 bp deletion resulted in the total elimination of *orf777*, while the 316 bp deletion affected the part of the *atp6* gene. Interestingly, in previous studies, we also discovered the removal of *orf777* from the mitochondrial genomes of two other CMS lines—HA89(PET2) [25] and HA89(MAX1) [26].

Table 2. Deletions (>100 bp) localized in the mitochondrial genome of HA89(ANN2) CMS line.

Deletion Length, bp	Positions in mtDNA of the Male-Fertile Line HA89	Deletion Localization according to the Male-Fertile Line HA89 Genetic Map	Deleted Genes
287	160,032–160,319	*rps13-nad6*	-
290	275,246–275,535	*atp6-cox2*	-
299	56,701–56,999	*nad4-ccmB*	-
316	268,901–269,216	*nad9-atp6*	*atp6* (partial)
583	70,338–70,920	*rpl10-nad1*	-
1165	35,574–36,738	*nad4L-orf777-atp8*	*orf777*
4204	190,339–194,542	*cob-ccmFC*	-
4575	29,197–33,771	*nad4L-orf777*	-
11,901	288,070–299,970	*cox2-nad2*	-

Seven longer than 100 bp insertions were detected in mtDNA of the HA89(ANN2) CMS line (Table 3). As a result of these insertions in the mitochondrial DNA of HA89(ANN2), five new open reading frames, namely *orf324*, *orf327*, *orf345*, *orf558*, and *orf933*, have appeared. All five ORFs are transcribed in the case of ANN2, contrary to the HA89 line.

A search for transmembrane domains (TDs) revealed that the protein encoded by *orf558* contained a single TD. In the case of *orf933*, two TDs were detected. The *orf933* encoded protein did not show homology to other sunflower proteins in GeneBank, and had only limited similarity (40–60 amino acids) to hypothetical mitochondrial proteins with unknown functions in *Lactuca sativa* (accession PLY70338.1), *Salvia miltiorrhiza* (accession YP_008992338.1), *Beta vulgaris* (accession CBJ23356.1), etc. Forty-six amino acids of the N-terminus of the protein encoded by *orf558* matched the N-terminus of cytochrome c oxidase subunit 2 (*cox2* gene). Moreover, most of the amino acids that form the transmembrane domain in *orf558* protein are identical to those in COX2. However, the sunflower cytochrome c oxidase subunit 2 has two TD and the protein encoded by *orf558*—only a single one (Figure 3). So the *orf558* represents a chimeric *cox2* gene and could potentially play a role in the ANN2 CMS phenotype development.

Table 3. Localization of insertions (>100 bp) in the mitochondrial genome of HA89(ANN2) CMS line.

Insertion in bp	Positions in mtDNA of HA89(ANN2)	New ORFs Based on Insertion	Homology to
430	147,733–148,162	*orf1197* *	*atp6*
1027	77,316–78,342	*orf324*	*orf285* (CMS PET2)
1310	149,412–150,722	*orf345*	
3757	137,947–141,703		
5338	295,554–300,891	*orf558*	*cox2*
6452	29,205–35,656		
9045	78,900–87,944	*orf327, orf933*	

* appeared as the result of several simultaneous reorganizations of mtDNA structure.

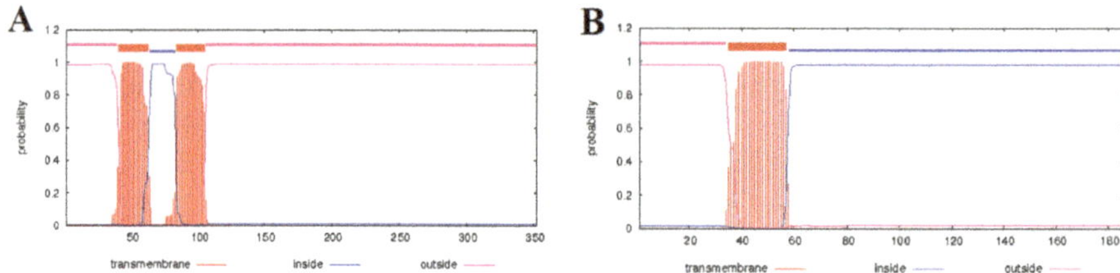

Figure 3. Comparison of transmembrane domains of proteins encoded by *cox2* (**A**) and *orf558* (**B**).

The most complex among the discovered ORFs in the HA89(ANN2) mitogenome was *orf1197*, which has appeared from three simultaneous reorganization events involving the 316 bp deletion, the 430 bp insertion, and the 6029 bp rearrangement. The *orf1197* represents a chimeric *atp6* gene, with transcription activity specific for the CMS line HA89(ANN2). In sunflower, the *atp6* gene typically encodes a protein consisting of 351 aa, whereas the predicted size of the translation product of the *orf1197* is 399 aa. Both proteins share 251 identical amino acids in the C-terminus. Thus, the protein encoded by the *orf1197* carries all seven TDs present in the C-terminus of the ATP synthase Fo subunit 6 from mitochondria of male-fertile sunflower (Figure 4).

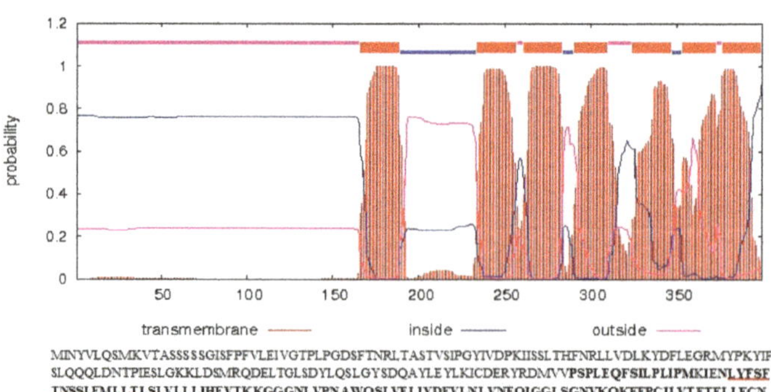

Figure 4. Comparison of *orf1197* and *atp6* encoded proteins and prediction of transmembrane helices. The amino acid sequence of the *orf1197* encoded protein is presented. Amino acids identical to ATP6 are shown in bold. Amino acids forming transmembrane domains are marked by red bars.

3. Discussion

Recently, we had investigated complete mitochondrial DNA sequences for three CMS sources in sunflower, coming from interspecific crosses: PET1, PET2, MAX1 [25,26]. Comparison of the HA89(ANN2) mitogenome with mitochondrial genome assemblies of male-fertile lines [25,27] and the other HA89 male-sterile analogs provides insights into reorganizations of mtDNA associated with CMS phenotypes. The male-fertile lines (HA89, HA412) have only slight variations in the mtDNA sequence [25]. Whereas the mitogenomes of the CMS sources (HA89(ANN2), HA89(PET1), (HA89(PET2), HA89(MAX1)) showed significant differences as compared to their alloplasmic male-fertile analog. The complete mitochondrial genome of the male-fertile line HA89 adds up to 300,947 bp (NCBI accession MN171345.1), while HA89(PET1) has a size of 305,217 bp (NCBI accession MG735191.1), HA89(PET2) of 316,582 bp (NCBI accession MG770607.2), HA89(MAX1) of 295,586 bp (NCBI accession MH704580.1) and HA89(ANN2) of 306,018 bp (MN175741.1). The difference in genomes sizes is due to several deletions and insertions. For instance, in the mtDNA of all the investigated CMS sources, except HA89(PET1), a similar deletion in the *nad4L-orf777-atp8* region was observed. In the case of HA89(PET2), this is due to a 711 bp deletion, in HA89(MAX1) has a 978 bp deletion, and in HA89(ANN2) there is an 1195 bp deletion. All these deletions resulted in removal of *orf777* from the mtDNA in the CMS lines. Another region enriched by deletions is the area between *cob-ccmFC*, here three overlapping deletions were detected: a 451 bp deletion in HA89(PET1), one of 3780 bp in HA89(PET2) and another one of 4204 bp in HA89(ANN2).

The coincidence between locations of these deletions is not accidental. There are three 265 bp repeats present in the sunflower mitochondrial genome, with the following positions in the mtDNA of the male-fertile HA89 line: 36537-36801 (adjacent to *atp8*), 190074-190338 (next to *cob*), and 202902-202638 (between *orf873-atp1*). These repeat regions are shown by red stars in Figure 5. Repeats represent common recombination points in mtDNA molecules [4,28]. The identification of small repeats involved in recombination is important because they influence the maintenance and evolution of mitochondrial genomes [28]. An imbalance in the nuclear-mitochondrion relationship that may occur in distant hybridizations impairs the recombination of mtDNA sub-genomic molecules, therefore, leading to reorganizations in the mitochondrial master chromosome. For instance, in HA89(PET1) the deletion, insertion, and inversion were mentioned in the *cob-atp1* region, directly between two repeats (Figure 5). In HA89(PET2), there were also several rearrangements in hot spots, resulting in the formation of a new gene cluster *cob-atp8-cox3*, as well as in the translocation of the *ccmFC-orf873-atp1* gene cluster into the *nad4L-orf777* region, combined with a deletion and huge insertion (Figure 5). In the mtDNA of both lines, HA89(MAX1) and HA89(ANN2), the specific *atp1-atp8-cox3* gene order was created, while in the MAX1 CMS source *ccmFC-orf873* translocated into the *nad4L-orf777* region (with deletion and huge insertion in this region). In the case of ANN2, the *ccmFC-orf873* region is located next to the *cob* gene (Figure 5). Thus, we have established that these three 265 bp repeats represent a reorganization hot spots in the sunflower mitochondrial genome.

Considering the insertions discovered in the HA89(ANN2) mitogenome, we also observed a similarity between the insertions of different CMS sources in sunflower. For instance, about 85% of the 1027 bp insertion sequence (ANN2) is complementary to the part of the 15,885 bp insertion (PET2). The 3757 bp insertion (ANN2) contains 1959 nucleotides identical to the 15,885 bp insertion (PET2) and 1215—to the 5272 bp insertion (MAX1). Also, 2343 bp of the 5338 bp insertion in ANN2 are similar to another region of sunflower mtDNA proximal to the *cob* gene (position 185,987–188,330 in the mtDNA of HA89 fertile line). So, this sequence is duplicated in the mitochondrial genome of the HA89(ANN2) CMS line. About 10% of 6452 bp insertion (ANN2) is complementary to both the 5050 bp (PET2) and the 5272 bp (MAX1) insertions. As well as 1158 bp of the 9044 bp insertion (ANN2) are identical to the 5050 bp and 15,885 bp insertions (PET2) and 5272 bp insertion (MAX1).

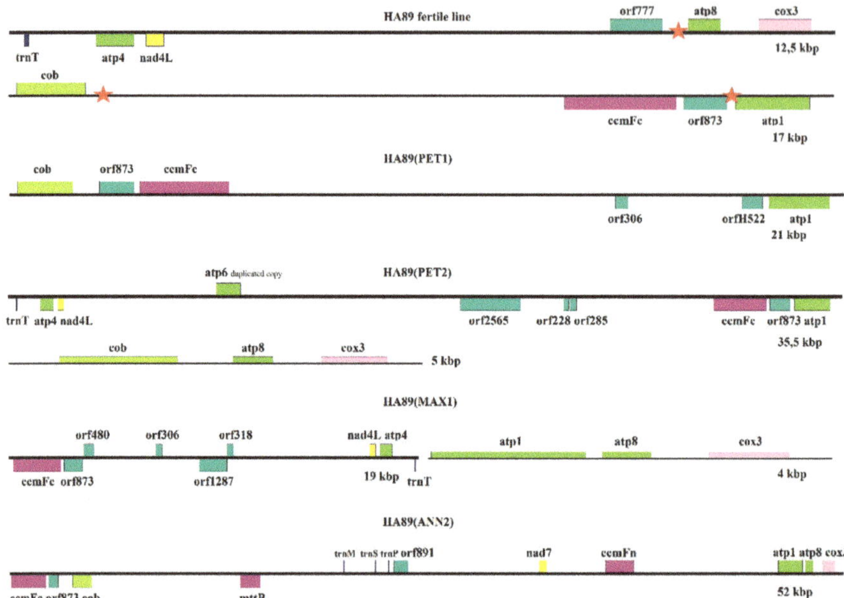

Figure 5. Reorganizations in CMS lines involving the 265 bp repeats (red stars) shown in the male-fertile mtDNA of HA89.

Although there are similarities in deletions, insertions, and rearrangements between mitochondrial genomes of the HA89(ANN2) and other CMS lines, the discovered ORFs were different. We summarized the data of all identified ORFs in the male-sterile cytoplasms in Table 4. The ORFs encoding proteins with similarity to other mitochondrial proteins and especially having transmembrane domains are of particular interest since the chimeric proteins with TD most often cause CMS phenotypes [10,29]. In the mtDNA of HA89(ANN2), we detected three new transcriptionally active ORFs, encoding proteins with TD—*orf558* (one TD), *orf933* (two TD), *orf1197* (seven TD). The *orf933* shows no homology to other sunflower genes, while *orf558* represents a chimeric *cox2* gene, and *orf1197* a chimeric *atp6* gene. It is difficult to estimate the exact contribution of *orf558*, *orf933*, *orf1197* to the development of the male-sterile phenotype in ANN2. However, the possibility of involvement of more than one open reading frame might be one explanation that ANN2 is so difficult to restore. On the other hand, the presence of a suppressor gene S1 discovered by Liu et al. [23] might be the reason for low rates of fertility restoration of the ANN2 CMS source. Previous studies indicate that the chimeric *atp6* genes or new ORFs that are co-transcribed with *atp6* most often cause CMS phenotypes in flowering plants [10,30–33]. Therefore, we suggest that *orf1197* is the major CMS candidate gene for ANN2 CMS source. Moreover, chimeric *atp6* genes were also identified in MAX1 [26] and CMS3/ANT1 [34] CMS types of sunflower. In CMS lines, chimeric *atp6* genes encode N-terminal extended proteins compared to the normal ATP synthase subunit 6 (351 aa): ANN2—399 aa, MAX1—429 aa, (AYV91168.1), CMS3/ANT1—437 aa (CAA57790.1). Moreover, 397 of 399 amino acids in the *orf1197* protein are identical to the chimeric *ATP6* of CMS3/ANT1 line, and therefore this protein represents a shorter version of this 437 aa-long protein. Such similarities support our hypothesis about the importance of *orf1197* in shaping the CMS phenotype in ANN2.

Table 4. Summary of transcriptionally active open reading frames in the mitochondrial genome of isonuclear fertile and CMS lines.

Fertile mtDNA	CMS ANN2	CMS MAX1	CMS PET2	CMS PET1
-	-	-	*orf228*	-
-	-	-	*orf285*	-
-	-	*orf306*	-	*orf306*
-	*orf324*	-	-	-
-	*orf327*	-	-	-
-	*orf345*	-	-	-
-	-	*orf480*	-	-
-	-	-	-	**orfH522**
-	*orf558*	-	-	-
-	-	*orf645*	*orf645*	-
orf777	-	-	-	*orf777*
-	**orf891**	-	-	-
-	**orf933**	-	-	-
-	**orf1197**	-	-	-
-	-	**orf1287**	-	-
-	-	-	*orf2565*	-

In bold: ORFs encoding proteins with transmembrane domains.

The *orf558* (as the chimeric *cox2* gene) might also cause cytoplasmic male sterility in ANN2. In other plants species, modified *cox2* sequences seem to be involved in the male sterility. For instance, the CMS specific *pcf* gene in petunia is composed from sequences of the 5′ portion of *atp9*, segments of *cox2*, and a large region of unknown origin—*urfS* [35]. In wild beets, the CMS-associated *orf129* shows homology to the 5′ flanking and the coding sequence of *cox2* [36]. In the mitogenome of the inap CMS source of *Brassica napus*, which was created by a somatic hybridization with *Isatis indigotica*, a novel *cox2-2* gene was detected, which represents recombination of the *cox2* of woad and *cox2-2* of rapeseed [37].

Another unique feature of HA89(ANN2) mitogenome is the formation of *orf891*. According to the ORFs predictions, a 3′ elongation of the *nad6* gene (*orf891*) may occur. However, the cDNA analyses did not agree with the genomic data. Perhaps due to *nad6* mRNA editing instead of a 3′ elongated transcript, the shorter one is formed. Heteromorphism in *nad6* transcript length was also observed in *Mimulus guttatus x M. nasutus* hybrids with the CMS phenotype [38]. Both male-fertile and male-sterile hybrids have a single copy of the *nad6* gene, and the divergence in mRNA length was only observed in male-sterile plants [38].

4. Materials and Methods

4.1. Plant Material

The CMS line HA89(ANN2) of sunflower was obtained from the genetic collection of the N. I. Vavilov All-Russian Institute of Plant Genetic Resources (Saint Petersburg, Russia). The original source of the ANN2 male-sterile cytoplasm was obtained by Serieys in 1984 [18]. All sunflower lines were grown in regularly irrigated pots in the growth chamber KBWF 720 (Binder, Tuttlingen, Germany) with the following growth conditions: temperature—26 °C, humidity—70%, photoperiod—10/14 h (dark/light).

4.2. Mitochondrial DNA Extraction, NGS Library Preparation, and Sequencing

First, the organelle fraction from leaves of 14-days-old sunflower seedlings was isolated, as described by Makarenko et al. [39]. Such preparations significantly reduced the amount of nuclear DNA. Then DNA extraction was performed with PhytoSorb kit (Syntol, Moscow, Russia), according to the manufacturer's protocol. Equal amounts of DNA from seven plants were mixed, and we used 1 ng of DNA pull for the NGS library preparation step. The library was made with Nextera XT DNA Library Prep Kit (Illumina, Mountain View, CA, USA), following the guidelines of Illumina. The quality of the library was evaluated using Bioanalyzer 2100 (Agilent, Santa Clara, CA, USA). The library was quantified at the Qubit fluorimeter (Invitrogen, Carlsbad, CA, USA) and by qPCR, then diluted up to the concentration of 8 pM. Sequencing was performed on two different Illumina sequencing platforms: HiSeq 2000 using TruSeq SBS Kit v3-HS 200-cycles and MiSeq using MiSeq Reagent Kit v2 500-cycles (Illumina, Mountain View, CA, USA). A total number of 3,063,836 reads (100-bp paired) and 3,305,268 reads (250-bp paired) were generated.

4.3. Mitochondrial Genome Assembly and Annotation

Quality control of reads was done using FastQC (https://www.bioinformatics.babraham.ac.uk/projects/fastqc/). Trimming of adapter-derived and low quality (Q-score below 25) reads was performed with Trimmomatic software [40]. For contig generation, we used SPAdes Genome Assembler v.3.10.1 [41]. The whole mitochondrial genome was manually assembled using scaffolds based on high coverage (depths > 70) contigs (length > 1000 kbp) and available bridge contigs (length = 0.3–1 kbp). The genome assembly was validated by remapping the initially obtained reads using Bowtie 2 v.2.3.3 [42]. All observed rearrangements were verified by PCR analysis. For variant calling, we used samtools/bcftools software [43] and manually revised polymorphic sites using the IGV tool [44].

The mitochondrial genome was annotated with MITOFY [45], BLAST tool [46], and ORFfinder (https://www.ncbi.nlm.nih.gov/orffinder). Using GeSeq [47], we provided comparisons of our annotation with the current reference annotations (NCBI accessions NC_023337.1, CM007908.1) of the sunflower mitochondrial genome. Graphical genome maps were generated using the OGDRAW tool v.1.3.1 [48]. The prediction of transmembrane domains was made with TMHMM Server v.2.0 (http://www.cbs.dtu.dk/services/TMHMM-2.0/). The scheme of the bioinformatic pipeline is presented in Supplementary Figure S1.

4.4. Expression Analyses

RNA was extracted from leaves of seven 28-days-old sunflower plants using a guanidinium thiocyanate-phenol-chloroform based method with the ExtractRNA reagent kit (Evrogen, Moscow, Russia). The quality and concentration of the RNA were evaluated with the Qubit fluorimeter (Invitrogen, Carlsbad, CA, USA) and the NanoDrop 2000 spectrophotometer (Thermo Fisher Scientific, Waltham, MA, USA). Total RNA (0.5 µg) was treated with DNAse I (Thermo Fisher Scientific, Waltham, MA, USA), and then cDNA was synthesized using the MMLV RT kit (Evrogen, Moscow, Russia) with random primers. As a negative control for each DNAse treated mRNA sample, the same reverse transcription protocol was performed, but without the MMLV enzyme. The quantitative PCR was performed with EvaGreen based RT-PCR kit (Syntol, Moscow, Russia) on Rotor-Gene 6000 (Corbett Research, Mortlake, NSW, Australia). A summary of all primer sequences is given in Supplementary Table S1. The primers annealing temperature (Tm) was 60 °C.

5. Conclusions

Assembly of CMS ANN2 mitochondrial genome (HA89(ANN2) line) revealed several rearrangements, insertions, and deletions, as well as seven new open reading frames: *orf324*, *orf327*, *orf345*, *orf558*, *orf891*, *orf933*, and *orf1197*. Transcripts were detected for all new open reading frames in CMS ANN2, but not in the fertile cytoplasm. Only *orf558*, *orf891*, *orf933*, and *orf1197* encoded proteins that contained membrane domains, making them the most likely CMS candidate genes for the ANN2 source. Notably, *orf1197* represents a chimeric *atp6* gene and presumably plays a major role in the CMS phenotype development of ANN2. However, CMS ANN2 may be caused by simultaneous action of several candidate genes. Hot spots for rearrangements (265 bp repeats) were identified, and we propose that they influence the maintenance and evolution of the mitochondrial genome in sunflower.

Supplementary Materials: The following are available online at http://www.mdpi.com/2223-7747/8/11/439/s1, Suppl Table S1. The primers sets used for transcription activity analysis, Suppl Table S2. The protein coding genes annotated in HA89(ANN2) mitochondrial genome, Suppl Figure S1. Bioinformatic pipeline for mitochondrial genome analysis. Trimmed reads are available at https://doi.org/10.6084/m9.figshare.8945084.v1.

Author Contributions: Conceptualization, K.V.A. and V.A.G.; methodology, M.S.M. and I.V.K.; software, T.V.T. and M.D.L.; validation, M.S.M., K.V.A. and I.V.K.; formal analysis, M.S.M. and T.V.T.; investigation, M.S.M., K.V.A. and M.D.L.; resources, V.A.G.; data curation, M.S.M. and M.D.L.; writing—original draft preparation, M.S.M., T.V.T. and R.H.; writing—review and editing, M.S.M., T.V.T. and R.H.; visualization, R.H.; supervision, A.V.U. and R.H.; project administration, A.V.U. and V.A.G.; funding acquisition, A.V.U. and M.D.L.

Funding: The study was supported by the Ministry of Education and Science of Russia project no. 6.929.2017/4.6. The NGS sequencing was provided with the support of budgetary subsidy to IITP RAS (Laboratory of Plant Genomics 0053-2019-0005). Analytical work was carried out on the equipment of centers for collective use of Southern Federal University "High Technology".

Acknowledgments: We are grateful to the cooperators of Northern Crop Science Laboratory (Fargo, USA) for providing seeds of HA89 alloplasmic lines of sunflower to the genetic collection of the N. I. Vavilov Institute of Plant Genetic Resources. We thank reviewers and the editor for their suggestions and comments, which has helped to improve the manuscript, and Todd Lorenz (tlorenz@laverne.edu), Duane R. Chartier (drc@authentica.org) for editing assistance.

Conflicts of Interest: The authors declare no conflict of interest. The funders had no role in the design of the study; in the collection, analyses, or interpretation of data; in the writing of the manuscript, or in the decision to publish the results.

References

1. Yang, J.; Liu, G.; Zhao, N.; Chen, S.; Liu, D.; Ma, W.; Hu, Z.; Zhan, M. Comparative mitochondrial genome analysis reveals the evolutionary rearrangement mechanism in *Brassica*. *Plant Biol.* **2015**, *18*, 527–536. [CrossRef]

2. Wang, S.; Song, Q.; Li, S.; Hu, Z.; Dong, G.; Song, C.; Huang, H.; Liu, Y. Assembly of a Complete Mitogenome of *Chrysanthemum nankingense* Using Oxford Nanopore Long Reads and the Diversity and Evolution of Asteraceae Mitogenomes. *Genes* **2018**, *9*, 547. [CrossRef] [PubMed]

3. Liu, H.; Cui, P.; Zhan, K.; Lin, Q.; Zhuo, G.; Guo, X.; Ding, F.; Yang, W.; Liu, D.; Hu, S.; et al. Comparative analysis of mitochondrial genomes between a wheat K-type cytoplasmic male sterility (CMS) line and its maintainer line. *BMC Genom.* **2011**, *12*, 163. [CrossRef] [PubMed]

4. Storchova, H.; Stone, J.D.; Sloan, D.B.; Abeyawardana, O.A.J.; Muller, K.; Walterova, J.; Pazoutova, M. Homologous recombination changes the context of *Cytochrome b* transcription in the mitochondrial genome of *Silene vulgaris* KRA. *BMC Genom.* **2018**, *19*, 874. [CrossRef] [PubMed]

5. Marechal, A.; Brisson, N. Recombination and the maintenance of plant organelle genome stability. *New Phytol.* **2010**, *186*, 299–317. [CrossRef]

6. Gualberto, J.M.; Mileshina, D.; Wallet, C.; Niazi, A.K.; Weber-Lotfi, F.; Dietrich, A. The plant mitochondrial genome: Dynamics and maintenance. *Biochimie* **2013**, *100*, 107–120. [CrossRef]

7. Zervas, A.; Petersen, G.; Seberg, O. Mitochondrial genome evolution in parasitic plants. *BMC Evol. Biol.* **2019**, *19*, 87. [CrossRef]

8. Ding, B.; Hao, M.; Mei, D.; Zaman, Q.U.; Sang, S.; Wang, H.; Wang, W.; Fu, L.; Cheng, H.; Hu, Q. Transcriptome and Hormone Comparison of Three Cytoplasmic Male-sterile Systems in *Brassica napus*. *Int. J. Mol. Sci.* **2018**, *19*, 4022. [CrossRef]

9. Laser, K.D.; Lersten, N.R. Anatomy and cytology of microsporogenesis in cytoplasmic male sterile angiosperms. *Bot. Rev.* **1972**, *33*, 337–346. [CrossRef]

10. Horn, R.; Gupta, K.J.; Colombo, N. Mitochondrion role in molecular basis of cytoplasmic male sterility. *Mitochondrion* **2014**, *19*, 198–205. [CrossRef]

11. Chen, L.; Liu, Y.-G. Male sterility and fertility restoration in crops. *Annu. Rev. Plant Biol.* **2014**, *65*, 579–606. [CrossRef] [PubMed]

12. Wu, Z.; Hu, K.; Yan, M.; Song, L.; Wen, J.; Ma, C.; Shen, J.; Fu, T.; Yi, B.; Tu, J. Mitochondrial genome and transcriptome analysis of five alloplasmic male-sterile lines in *Brassica juncea*. *BMC Genom.* **2019**, *20*, 348. [CrossRef] [PubMed]

13. Chen, Z.; Zhao, N.; Li, S.; Grover, C.E.; Nie, H.; Wendel, J.F.; Hua, J. Plant Mitochondrial Genome Evolution and Cytoplasmic Male Sterility. *Crit. Rev. Plant Sci.* **2017**, *36*, 55–69. [CrossRef]

14. Serieys, H. Identification, study and utilisation in breeding programs of new CMS sources in the FAO subnetwork. In Proceedings of the 2005 Sunflower Subnetwork Progress Report, Novi Sad, Serbia and Montenegro, 17–20 July 2005; FAO: Rome, Italy, 2005; pp. 47–53.

15. Vear, F. Changes in sunflower breeding over the last fifty years. *OCL* **2016**, *23*, D202. [CrossRef]

16. Leclercq, P. Une sterilite male chez le tournesol. *Ann. Amelior. Plant.* **1969**, *19*, 99–106.

17. Horn, R.; Friedt, W. CMS sources in sunflower: Different origin but same mechanism? *Theor. Appl. Genet.* **1999**, *98*, 195–201. [CrossRef]

18. Serieys, H. Report on the activities of the F.A.O. working group: "Identification, study and utilization in breeding programs of new CMS sources", for the period 1991–1993. *HELIA* **1994**, *17*, 93–102.

19. Serieys, H.; Vincourt, P. Caracterisation de nouvelles sources de sterility male chez le tournesol. *Les Colloq. INRA Paris* **1987**, *45*, 53–64.

20. Horn, R.; Friedt, W. Fertility restoration of new CMS sources in sunflower (*Helianthus annum* L.). *Plant Breed.* **2006**, *116*, 317–322. [CrossRef]

21. Jan, C.C. Cytoplasmic male sterility in two wild *Helianthus annuus* L. accessions and their fertility restoration. *Crop Sci.* **2000**, *40*, 1535–1538. [CrossRef]

22. Jan, C.C. Silencing of fertility restoration genes in sunflower. *HELIA* **2003**, *26*, 1–6. [CrossRef]

23. Liu, Z.; Long, Y.; Xu, S.S.; Seiler, G.; Jan, C.C. Unique fertility restoration suppressor genes for male-sterile CMS ANN2 and CMS ANN3 cytoplasms in sunflower (*Helianthus annuus* L.). *Mol. Breed.* **2019**, *39*, 22. [CrossRef]

24. Horn, R. Molecular diversity of male sterility inducing and male-fertile cytoplasms in the genus *Helianthus*. *Theor. Appl. Genet.* **2002**, *104*, 562–570. [CrossRef] [PubMed]

25. Makarenko, M.; Kornienko, I.; Azarin, K.; Usatov, A.; Logacheva, M.; Markin, N.; Gavrilova, V. Mitochondrial genomes organization in alloplasmic lines of sunflower (*Helianthus annuus*) with various types of cytoplasmic male sterility. *PeerJ* **2018**, *6*, e5266. [CrossRef] [PubMed]

26. Makarenko, M.S.; Usatov, A.V.; Tatarinova, T.V.; Azarin, K.V.; Logacheva, M.D.; Gavrilova, V.A.; Horn, R. Characterization of the mitochondrial genome of the MAX1 type of cytoplasmic male-sterile sunflower. *BMC Plant Biol.* **2019**, *19*, 51. [CrossRef]

27. Grassa, C.J.; Ebert, D.P.; Kane, N.C.; Rieseberg, L.H. Complete Mitochondrial Genome Sequence of Sunflower (*Helianthus annuus* L.). *Genome Announc.* **2016**, *4*, e00981-16. [CrossRef]

28. Wynn, E.L.; Christensen, A.C. Repeats of Unusual Size in Plant Mitochondrial Genomes: Identification, Incidence and Evolution. *G3 Genes Genomes Genet.* **2019**, *9*, 549–559. [CrossRef]

29. Mower, J.P.; Case, A.L.; Floro, E.R.; Willis, J.H. Evidence against Equimolarity of Large Repeat Arrangements and a Predominant Master Circle Structure of the Mitochondrial Genome from a Monkeyflower (*Mimulus guttatus*) Lineage with Cryptic CMS. *Genome Biol. Evol.* **2012**, *4*, 670–686. [CrossRef]

30. Yamamoto, M.P.; Kubo, T.; Mikami, T. The 5′-leader sequence of sugar beet mitochondrial atp6 encodes a novel polypeptide that is characteristic of Owen cytoplasmic male sterility. *MGG* **2005**, *273*, 342–349. [CrossRef]

31. Kim, D.H.; Kim, B.D. The organization of mitochondrial atp6 gene region in male-fertile and CMS lines of pepper (*Capsicum annuum* L.). *Curr. Genet.* **2006**, *49*, 59–67. [CrossRef]

32. Jing, B.; Heng, S.; Tong, D.; Wan, Z.; Fu, T.; Tu, J.; Ma, C.; Yi, B.; Wen, J.; Shen, J. A male sterility-associated cytotoxic protein ORF288 in *Brassica juncea* causes aborted pollen development. *J. Exp. Bot.* **2012**, *63*, 1285–1295. [CrossRef] [PubMed]

33. Tan, G.F.; Wang, F.; Zhang, X.Y.; Xiong, A.I. Different lengths, copies and expression levels of the mitochondrial atp6 gene in male-sterile and fertile lines of carrot (*Daucus carota* L.). *Mitochondrial DNA Part A* **2017**, *29*, 446–454. [CrossRef] [PubMed]

34. Spassova, M.; Moneger, F.; Leaver, C.J.; Petrov, P.; Atanassov, A.; Nijkamp, H.J.; Hille, J. Characterisation and expression of the mitochondrial genome of a new type of cytoplasmic male-sterile sunflower. *Plant Mol. Biol.* **1994**, *26*, 1819–1831. [CrossRef] [PubMed]

35. Young, E.G.; Hanson, M.R. A fused mitochondrial gene associated with cytoplasmic male sterility is developmentally regulated. *Cell* **1987**, *50*, 41–49. [CrossRef]

36. Yamamoto, M.P.; Shinada, H.; Onodera, Y.; Komaki, C.; Mikami, T.; Kubo, T. A male sterility-associated mitochondrial protein in wild beets causes pollen distribution in transgenic plants. *Plant J.* **2008**, *54*, 1027–1036. [CrossRef]

37. Kang, L.; Li, P.; Wang, A.; Ge, X.; Li, Z. A Novel Cytoplasmic Male Sterility in *Brassica napus* (inap CMS) with Carpelloid Stamens via Protoplast Fusion with Chinese Woad. *Front. Plant Sci.* **2017**, *8*, 529. [CrossRef]

38. Case, A.L.; Willis, J.H. Hybrid male sterility in *Mimulus* (*Phrymaceae*) is associated with a geographically restricted mitochondrial rearrangement. *Evolution* **2008**, *62*, 1026–1039. [CrossRef]

39. Makarenko, M.S.; Usatov, A.V.; Markin, N.V.; Azarin, K.V.; Gorbachenko, O.F.; Usatov, N.A. Comparative Genomics of Domesticated and Wild Sunflower: Complete Chloroplast and Mitochondrial Genomes. *OnLine J. Biol. Sci.* **2016**, *16*, 71–75. [CrossRef]

40. Bolger, A.M.; Lohse, M.; Usadel, B. Trimmomatic: A flexible trimmer for Illumina sequence data. *Bioinformatics* **2014**, *30*, 2114–2120. [CrossRef]

41. Nurk, S.; Bankevich, A.; Antipov, D.; Gurevich, A.; Korobeynikov, A.; Lapidus, A.; Prjibelsky, A.; Pyshkin, A.; Sirotkin, A.; Sirotkin, Y. Assembling Genomes and Mini-metagenomes from Highly Chimeric Reads. In *Research in Computational Molecular Biology*; Deng, M., Jiang, R., Sun, F., Zhang, X., Eds.; RECOMB 2013; Lecture Notes in Computer Science; Springer: Berlin/Heidelberg, Germany, 2013; Volume 7821, pp. 158–170. [CrossRef]

42. Langmead, B.; Salzberg, S. Fast gapped-read alignment with Bowtie 2. *Nat. Methods* **2012**, *9*, 357–359. [CrossRef]

43. Li, H. A statistical framework for SNP calling, mutation discovery, association mapping and population genetical parameter estimation from sequencing data. *Bioinformatics* **2011**, *27*, 2987–2993. [CrossRef] [PubMed]

44. Thorvaldsdóttir, H.; Robinson, J.T.; Mesirov, J.P. Integrative Genomics Viewer (IGV): High-performance genomics data visualization and exploration. *Brief. Bioinform.* **2013**, *14*, 178–192. [CrossRef] [PubMed]

45. Alverson, A.J.; Wei, X.; Rice, D.W.; Stern, D.B.; Barry, K.; Palmer, J.D. Insights into the Evolution of Mitochondrial Genome Size from Complete Sequences of *Citrullus lanatus* and *Cucurbita pepo* (Cucurbitaceae). *Mol. Biol. Evol.* **2010**, *27*, 1436–1448. [CrossRef] [PubMed]

46. Altschul, S.F.; Gish, W.; Miller, W.; Myers, E.W.; Lipman, D.J. Basic local alignment search tool. *J. Mol. Biol.* **1990**, *215*, 403–410. [CrossRef]

47. Tillich, M.; Lehwark, P.; Pellizzer, T.; Ulbricht-Jones, E.S.; Fischer, A.; Bock, R.; Greiner, S. GeSeq—Versatile and accurate annotation of organelle genomes. *Nucleic Acids Res.* **2017**, *45*, W6–W11. [CrossRef] [PubMed]

48. Greiner, S.; Lehwark, P.; Bock, R. OrganellarGenomeDRAW (OGDRAW) version 1.3.1: Expanded toolkit for the graphical visualization of organellar genomes. *Nucleic Acids Res.* **2019**, *47*, W59–W64. [CrossRef]

Article

The First Plastid Genome of the Holoparasitic Genus *Prosopanche* (Hydnoraceae)

Matthias Jost [1], Julia Naumann [1], Nicolás Rocamundi [2], Andrea A. Cocucci [2] and Stefan Wanke [1,*

[1]   Technische Universität Dresden, Institut für Botanik, 01062 Dresden, Germany; matthias.jost@tu-dresden.de (M.J.); Julia.naumann@tu-dresden.de (J.N.)
[2]   Instituto Multidisciplinario de Biología Vegetal, Universidad Nacional de Córdoba, CONICET, FCEFyN, Av. Vélez Sársfield 1611, Córdoba X5016GCA, Argentina; nicolasrocamundi@gmail.com (N.R.); aacocucci@imbiv.unc.edu.ar (A.A.C.)
*   Correspondence: stefan.wanke@tu-dresden.de; Tel.: +49-351-463-34281

Received: 11 January 2020; Accepted: 11 February 2020; Published: 1 March 2020

Abstract: Plastomes of parasitic and mycoheterotrophic plants show different degrees of reduction depending on the plants' level of heterotrophy and host dependence in comparison to photoautotrophic sister species, and the amount of time since heterotrophic dependence was established. In all but the most recent heterotrophic lineages, this reduction involves substantial decrease in genome size and gene content and sometimes alterations of genome structure. Here, we present the first plastid genome of the holoparasitic genus *Prosopanche*, which shows clear signs of functionality. The plastome of *Prosopanche americana* has a length of 28,191 bp and contains only 24 unique genes, i.e., 14 ribosomal protein genes, four ribosomal RNA genes, five genes coding for tRNAs and three genes with other or unknown function (*accD*, *ycf*1, *ycf*2). The inverted repeat has been lost. Despite the split of *Prosopanche* and *Hydnora* about 54 MYA ago, the level of genome reduction is strikingly congruent between the two holoparasites although highly dissimilar nucleotide sequences are observed. Our results lead to two possible evolutionary scenarios that will be tested in the future with a larger sampling: 1) a Hydnoraceae plastome, similar to those of *Hydnora* and *Prosopanche* today, existed already in the most recent common ancestor and has not changed much with respect to gene content and structure, or 2) the genome similarities we observe today are the result of two independent evolutionary trajectories leading to almost the same endpoint. The first hypothesis would be most parsimonious whereas the second would point to taxon dependent essential gene sets for plants released from photosynthetic constraints.

Keywords: Piperales; Hydnoraceae; *Hydnora*; *Prosopanche*; parasitic plants; holoparasite; plastid genome

---

1. Introduction

In photoautotrophic plants, the plastid chromosome encodes essential genes for major photosynthesis related functions and is mostly considered highly conserved with respect to gene order and gene content [1] as well as its organization as quadripartite structure consisting of two single copy regions separated by two copies of an inverted repeat [2,3]. The plastomes of parasitic plants are a prime subject to study the possibilities and changes of otherwise highly conserved genomic structures that can occur when organellar genomes are released from selective constraints [4,5]. Plastid genomes of members from nearly all of the at least 11 independently evolved parasitic angiosperm lineages [6] have been studied to a more or less extensive degree [3,7–15]. Recent in-depth papers and reviews summarized similar evolutionary stages of plastome decay [5,16,17], eventually resulting in possible complete loss in Rafflesiaceae [14], depending on the respective level of heterotrophy. However, a vast variety of changes in plastid genome size and organization, gene content, nucleotide composition and mutational rates has been reported within individual lineages and the speed and extent of degradation seem lineage-specific [3]. Despite those lineage-specific differences, an underlying pattern can be found that plants follow as they transition from autotrophic to heterotrophic lifestyle. With increasing heterotrophy, genes involved in photosynthesis show signatures of relaxed functional constraints and eventually become pseudogenized and lost, often starting with *ndh* genes [16,18]. Following the initial losses is a relaxation of purifying selection in genes related to photosynthesis and genes involved in the translation and transcription machinery. General housekeeping genes that carry out non-photosynthesis related functions can show less relaxation or even slightly higher purification [18]. There are many examples of the different stages of parasitism in numerous lineages with the Orobanchaceae being the prime subject to study the whole range from facultative to holoparasitism [19,20]. It is apparent that often gene order and plastome structure such as the quadripartite nature (large single copy region LSC, small single copy region SSC and inverted repeats IR) are retained in different parasitic plastomes. Yet, there are some exceptions such as the loss of the inverted repeat for example in the highly reduced plastomes of *Pilostyles* [12,21], *Balanophora* [22] and many other lineages of autotrophic plants [23–26]. Functionally the inverted repeats are hypothesized to stabilize the plastid genome [27]; therefore, their loss is shown to lead to more frequent rearrangements [28].

One of the 11 parasitic angiosperm lineages is the holoparasitic family Hydnoraceae (Piperales). In contrast to all other parasitic angiosperm lineages Hydnoraceae are the only lineage outside the monocot and eudicot radiation and among one of the oldest parasitic lineages with an estimated stem group age of ~91 MYA [29]. The family consists of the two genera *Hydnora* (7 species) [30] and *Prosopanche* (5/6 species) [31] with *Hydnora* occurring exclusively in the Old World and *Prosopanche* in the New World [32,33]. According to molecular dating analyses the two genera split from each other about 54 MYA [29]. The plastid genomes of Hydnoraceae are also among the least known with a single known plastome of *Hydnora visseri* [10], for which not only a drastic reduction in genome size but also in gene content has been shown, exceeding the loss of genes involved in the photosynthesis apparatus, yet maintaining the quadripartite structure of the plastome as well as showing nearly identical gene order and orientation compared to the closely related photoautotrophic genus *Piper* (Piperaceae, Piperales) [10]. Given that the two Hydnoraceae genera diverged in the early Eocene and the plastome of *Hydnora visseri* is among the most reduced holoparasitic angiosperms knowledge about its sister genus *Prosopanche* would be highly valuable for putting the sequenced plastomes in context to better understand the evolution of this specific lineage and to compare it to the other different parasitic lineages. We here report the plastid genome of *Prosopanche americana* and compare it to the published genome of *Hydnora visseri* as well as to the plastome of *Aristolochia contorta* [34] (Aristolochiaceae, the closest autotrophic relatives of Hydnoraceae).

2. Results

2.1. The Plastome of Prosopanche Americana

In total, 373 million reads were sequenced and de novo assembled using CLC Workbench [35], resulting in a total of 8,342,403 scaffolds of which 372 had BLAST hits for plastid features (Figure 1), using a query of 45 angiosperm plastomes from GenBank and setting the e-value to 1e-10. In depth analyses of these scaffolds based on quality of BLASTn hit in combination with scaffold coverage, resulted in the exclusion of all but scaffold 424, which received the highest number of BLAST hits for plastid genome features. Additionally, it is the longest scaffold with plastid BLAST hits (28,191 bp) and shows the highest depth of coverage (4678) and with a total of 952,705 reads mapped to it. The scaffold has identical ends in sequence, forming 47 bp of overlap, allowing for circularization. The circularization was verified by PCR and the Sanger sequenced product was aligned to the assembled scaffold, which also revealed and corrected a misassembly and an insertion at the respective contig ends, but introduces two unresolved characters in this region.

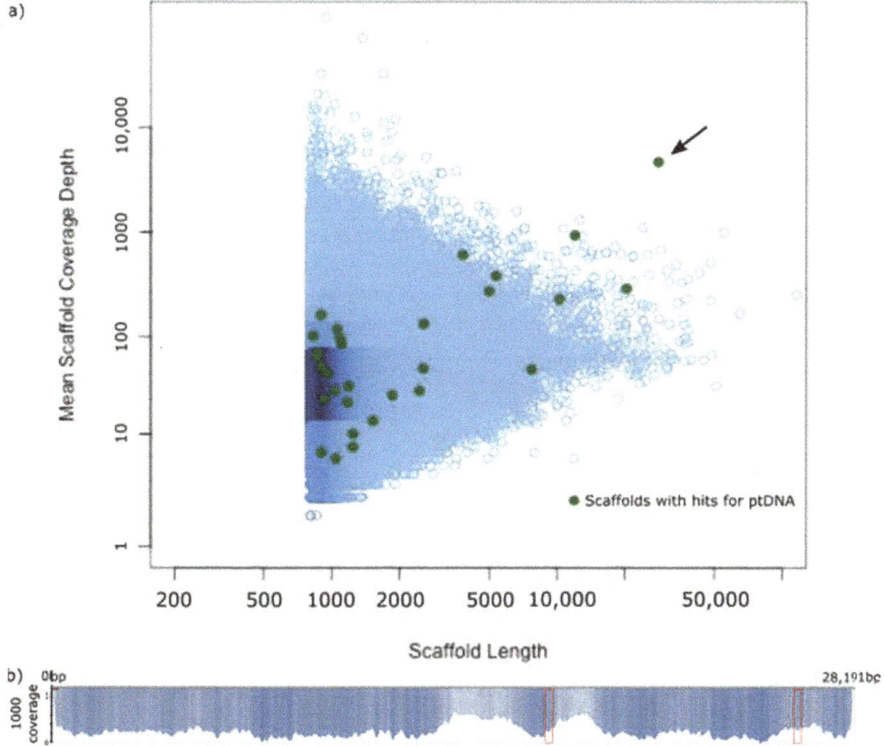

Figure 1. The plastid genome of *Prosopanche americana* is found on a single scaffold. **a)** Stoichiometry plot of the scaffold lengths and the respective mean coverage depth of the *Prosopanche americana* assembly. Scaffolds are visualized as blue circles. The green dots represent scaffolds with BLAST hits for annotated fractions of the plastome (derived from "gene features" in NCBI). The black arrow points to scaffold 424. **b)** Distribution of the reads mapped to the linear scaffold 424 with the scale displaying the depth of coverage in 1k increments. The red boxes represent the coverage of the *Prosopanche amercana* repeat regions.

With this, the *Prosopanche americana* plastome is 28,191 bp in length (Figure 2). Using DOGMA (% identity cutoff = 25, e-value = 1e-5) [36], 25 potential plastid genes were initially identified, consisting of 17 protein coding genes, 3 rRNAs and 7 tRNAs. Only 24 of the initial 25 genes could be verified and were used for further testing and analysis because the hit for *trn*P-GGG could not be confirmed and is a result of the low stringency search settings. With the methods described below, one additional rRNA could be identified, which was not detected by the DOGMA analysis. For comparison, we also ran an analysis of scaffold 424 using MFannot [37] which resulted in a similar result, but failed to identify *rrn*5, *ycf*2 or *trn*fM-CAU. *trn*W identified by DOGMA is identified as *trn*C with MFannot. In total, the plastome contains 25 genes, of which 24 are unique. There are 14 ribosomal protein genes (*rpl*2, *rpl*14, *rpl*16, *rpl*36, *rps*2, *rps*3, *rps*4, *rps*7, *rps*8, *rps*11, *rps*12, *rps*14, *rps*18, *rps*19), two of which contain introns (*rpl*16, *rps*12), four ribosomal RNA genes (*rrn*4.5, *rrn*5, *rrn*16, *rrn*23), five genes coding for tRNAs (*trn*E-UUC, 2x *trn*I-CAU, *trn*fM-CAU, *trn*W-CCA, *trn*Y-GUA) and three genes with other or unknown functions (*acc*D, *ycf*1, *ycf*2). The *Prosopanche americana* plastome (GenBank accession number MT075717) lacks genes involved in photosynthesis, such as genes for the photosystems I and II, RuBisCO large subunit, cytochrome-related or ATP synthase, NADH dehydrogenase genes, as well as RNA polymerase subunits, *clp*P and *mat*K. All protein coding genes in the plastome have feature continuous reading frames with the only exception of the *rps*19 gene, which does not possess an in-frame start codon in close proximity. Amino acid isotypes predicted by DOGMA [36] are confirmed by tRNAscan-SE [38,39] anticodon prediction (Figure S1) in all but the case of *trn*W-CCA. The underlying sequence is predicted to form a cloverleaf structure with three leaves but it has an ACA anticodon (as predicted by MFannot) instead of CCA. Alignments of the respective tRNAs from *Aristolochia* (*Aristolochia contorta* NC_036152) [34] indicate that this tRNA is homologous to the *Aristolochia trn*W-CCA. A three-loop cloverleaf 2D structure is predicted for all identified tRNAs, except for *trn*Y-GUA. The latter shows an additional, variable arm in the tRNAscan-SE prediction, which is also present in the *trn*Y of *Aristolochia contorta*. The anticodon versus isotype model prediction (Table S1) by tRNAscan-SE is consistent for two tRNAs (*trn*fM-CAU and *trn*Y-GUA). The correctness of the anticodon prediction for all tRNAs was verified via alignments with respective genes of *Aristolochia contorta*. Given the abovementioned information, it is likely that *trn*W-CCA is a pseudogene.

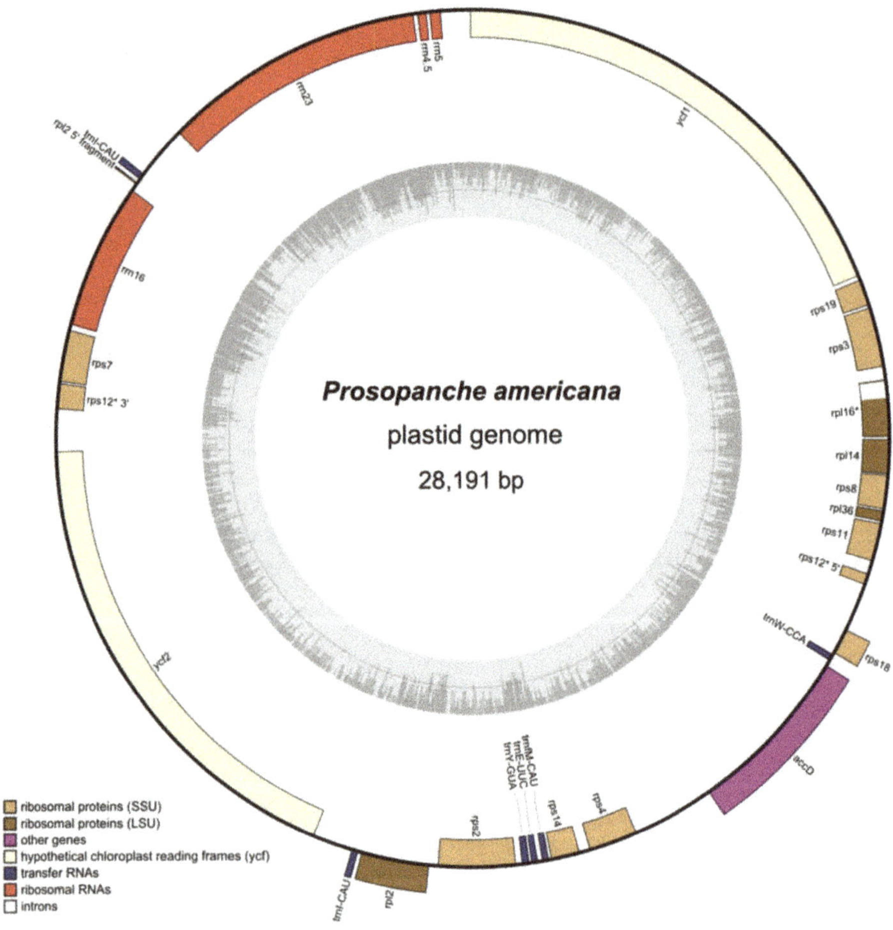

Figure 2. The highly reduced plastome of *Prosopanche americana*. The circular plastome of holoparasitic *Prosopanche americana* contains 24 unique genes (14 ribosomal protein genes, 4 ribosomal RNA genes, five genes coding for transfer RNAs and three genes with other or unknown function), which are distributed over a total length of 28,191 bp. Gene distribution is displayed on the outer circle with color coded gene groups according to the legend (bottom left). GC content is visualized as inner, grey circle. * indicates genes with introns.

The *Prosopanche* plastome contains 27 intergenic regions with an average length of 173 bp and the longest one being 879 bp in length. In total, 23 open reading frames were identified within the intergenic regions (ORF finder, implemented in Geneious 11.1.5 [40]), while setting the minimum size to 90 bp, which is the shortest plastid encoded protein coding gene in Piperales and allowing for the detection of smaller ORFs within a larger ORF. In-depth analyses of intergenic regions through BLAST or automatic annotation (as low as 50% sequence similarity) did not return meaningful hits for genes or pseudogenes. Structurally, the plastome of *Prosopanche americana* does not have any inverted repeats and therefore does not show the typical quadripartite structure. All genes are found as a single copy on the plastome, except the *trn*I-CAU gene, which is found in two copies both showing the same orientation on the circular plastome. Analysis of the flanking regions of both copies (Figure 3) show a stretch of 195 bp with only four substitutions between them. In addition to the fully duplicated tRNA, an alignment of the two copies also shows a duplication of the first 25 bp of the *rpl*2 gene. The gene is truncated downstream in one repeat (copy B) due to sequence divergence and the end of the duplication. The coverages of the duplication copies are neither significantly higher nor significantly lower than the average plastome coverage (4678). In copy A coverage ranges from 4285 to 4682 and from 4316 to 4868 for copy B.

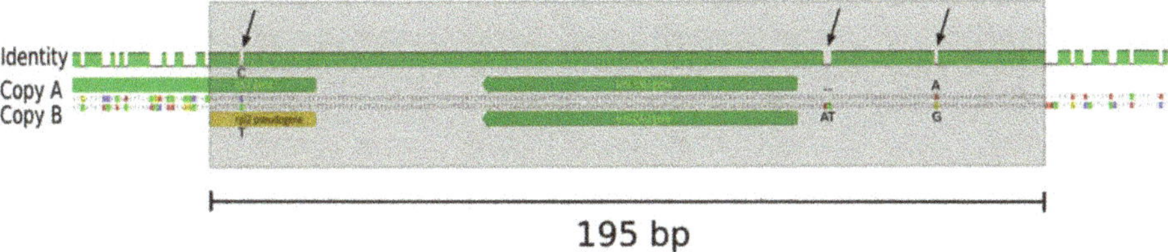

Figure 3. *Prosopanche americana trn*I-CAU repeat region. Alignment of the two *trn*I-CAU repeat copies highlights high sequence identity (green identity bar) over 195 bp with only four nucleotide differences (arrows pointing to gaps in identity graph and nucleotide differences highlighted by color and nucleotide code in sequence alignment). The *trn*I-CAU gene is fully contained within the repeat whereas only 25 bp of the *rpl*2 gene start fall within repeat borders. Contrary to copy B, which stays with a truncated *rpl*2 pseudogene, copy A is preceded by the complete *rpl*2 ORF of *Prosopanche americana*.

The GC content of the whole plastome is 20.4% (Figure 2, Figure 4a). Protein coding regions, which make up 64.9% (18,303 bp) of the total length consist to 17.9% of GC nucleotides, rRNA genes make up 16.5% (4639 bp) and contain with 37.2% the highest percentage of GC in the *Prosopanche* plastome (Figures 2 and 4a). Noncoding regions (excluding introns) on the other hand show the least amount of GC content with 11.3%, while making up 16.3% (4600 bp) of the plastome.

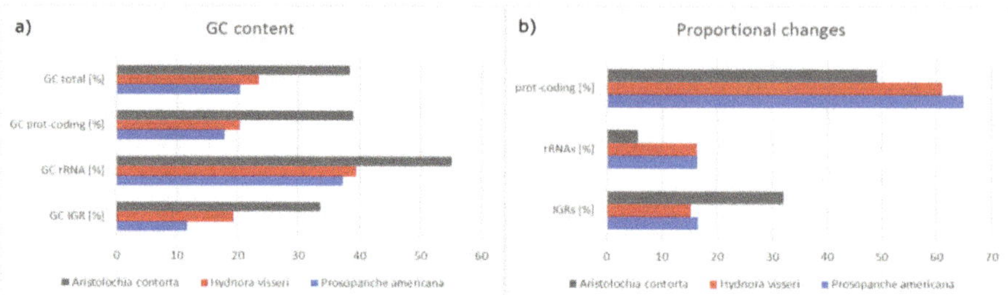

Figure 4. *Prosopanche americana* has a low GC content. (**a**) Comparison of the GC content and (**b**) proportional changes of plastid genome compartments (protein coding, ribosomal RNA, intergenic region) among the photosynthetic *Aristolochia contorta* (Aristolochiaceae, NC_036152) and the holoparasitic Hydnoraceae highlighting the extent of changes the plastomes of the Hydnoraceae have undergone with respect to nucleotide composition and conservation of specific genome regions.

2.2. Comparison of the Plastomes of the Holoparasitic Hydnoraceae

The plastome of *Prosopanche americana* (28,191 bp) is slightly larger than the plastome of *Hydnora visseri* (27,233 bp, NC_029358) even though it lacks the inverted repeat, which contributes 1466 bp to the length of *Hydnora*. *Prosopanche*, however, contains a 195 bp repeat and two additional genes (Figure 5). The remaining set of genes is shared between the genomes of *Hydnora* and *Prosopanche*. All protein coding genes are putatively functional with the exception of the *Prosopanche rps*19 and the *Hydnora rps*7. Single gene alignment of the *accD* gene shows a drastic size difference between the copies of the two Hydnoraceae genera. The closest ATG start codon in frame eligible as *accD* gene start of *Prosopanche* leads to a 184 bp increase in length, which does not align to the *Hydnora accD* gene. Apart from the protein coding genes, the set of ribosomal RNAs also is identical between the sister taxa, solely the number of transfer RNAs differs between the two plastomes. *Prosopanche* contains with *trn*W and *trn*Y two additional tRNAs, yet the former in the form of a pseudogene. The gene order between the two plastomes is mostly consistent with an exception of an inversion of the '*rps*7-*rps*12 exon 2 and 3'- region and the '*trn*I-*rpl*2-*rps*2'- region. Contrary to *Hydnora visseri*, the plastome of *Prosopanche americana* does not possess inverted repeats, though it contains a duplicated *trn*I-CAU-region while both copies share the same orientation. The same tRNA, *trn*I, is the only complete gene duplicated in the inverted repeat of *Hydnora*. The overall GC content in *Prosopanche* is lower than in its sister genus (Figure 4a) with the most drastic difference showing in the non-coding regions. The dotplot (Figure A1 in Appendix A) of the two plastomes shows a fairly straight diagonal line with many interruptions due to differences on sequence level and an inversion towards the end (Figure A1, red circle) illustrating the larger of the two inversions between the species.

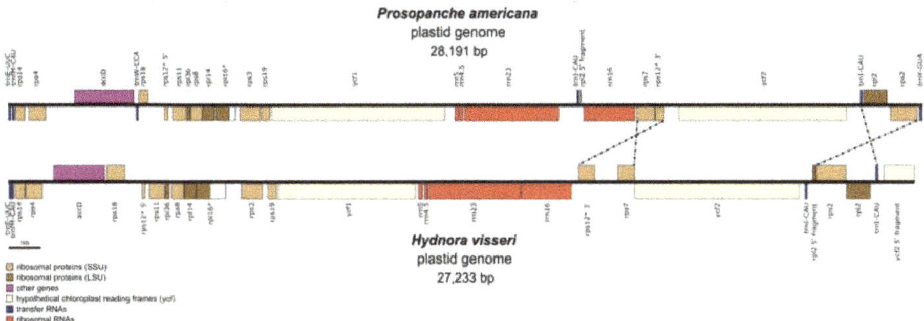

Figure 5. High structural similarity between the plastomes of Hydnoraceae. Linear plastid genome comparison of *Prosopanche americana* and *Hydnora visseri* (NC_029358). Genes are color coded; the dotted lines highlight inversions between the two genomes compared, genes containing introns are marked with *.

2.3. Plastome Comparison to Aristolochia

A comparison of the plastomes of Hydnoraceae with the closely related photoautotrophic *Aristolochia contorta* (NC_036152) [34] demonstrates large differences in genome size, gene and GC content. With a length of 160,576 bp, the plastome of *A. contorta* is about six times the size of the holoparasites and with a total of 131 genes it contains about six times the amount of that of *Prosopanche americana* and *Hydnora visseri*. *Aristolochia contorta* shows a characteristic plastid genome for photoautotrophic Piperales, with all necessary genes for the photosynthesis machinery [34]. The overall GC content (38.3%), as well as the GC content in different parts of the plastome, is higher than in the respective parts of the Hydnoraceae (Figure 4a) with the highest overall GC contents found in the rRNA genes across the three genera. The reduction of G and C nucleotides in *Hydnora* and *Prosopanche* occurs at a similar rate, with the latter having slightly lower percentages than *Hydnora*. The intergenic regions of *P. americana* however, show a much more drastic increase in AT content than observed in *Hydnora* and make up to 88.3% of the nucleotides (Figure 4). The comparison of changes within the three categories protein coding genes, rRNAs and intergenic regions (IGR) highlights the proportional increase in length in relation to the total plastome length for protein coding genes and the four rRNA genes (Figure 4b). The rRNA genes of the Hydnoraceae make up about three times as much of the total plastome length compared to the corresponding genes in *A. contorta*. A proportional decrease is shown for the intergenic regions in the parasitic plastomes compared to the photoautotrophic relative where IGRs contribute to about twice as much to the total plastome length.

An in-depth look at GC content and codon usage of the protein coding genes shared between the three plastomes using CodonW [41] shows a decrease in G and C nucleotides at the third codon position of these genes in the Hydnoraceae (Table S2). The two protein coding genes with the lowest GC proportion in the parasitic plastomes are *ycf*2 and *rps*18, with *ycf*2 being also one of the longest plastid-encoded genes. The lowest GC content in *A. contorta* is found in the *ycf*1 gene with 31.9%, which is the highest percentage found in all of the *H. visseri* protein coding genes. The lowest overall G and C percentage is found in the *ycf*2 gene of *P. americana* with only 13.3%.

2.4. Phylogenetic Placement of Hydnoraceae

The phylogenetic maximum likelihood (ML) tree (Figure 6) of the concatenated 82 gene alignment estimated with GTR+I+G substitution model and rapid bootstrapping (1000 replicates) recovers all major clades as monophyletic with high backbone support, except for the eudicot-monocot split which has a lower bootstrap value of 68. *Prosopanche* and *Hydnora* are supported as sister genera (bootstrap value = 100) and the monophyletic Hydnoraceae are placed within the Piperales as sister to a sister group of Aristolochioideae and Asaroideae with low support (bootstrap value = 13). Piperales as a monophyletic group receive only moderate support (bootstrap value = 76). The phylogram (Figure 6, left) highlights extremely long branches, not only leading to the most recent common ancestor (MRCA) of *Hydnora* and *Prosopanche* but also from the MRCA to the terminals, demonstrating the extremely high substitution rates in Hydnoraceae.

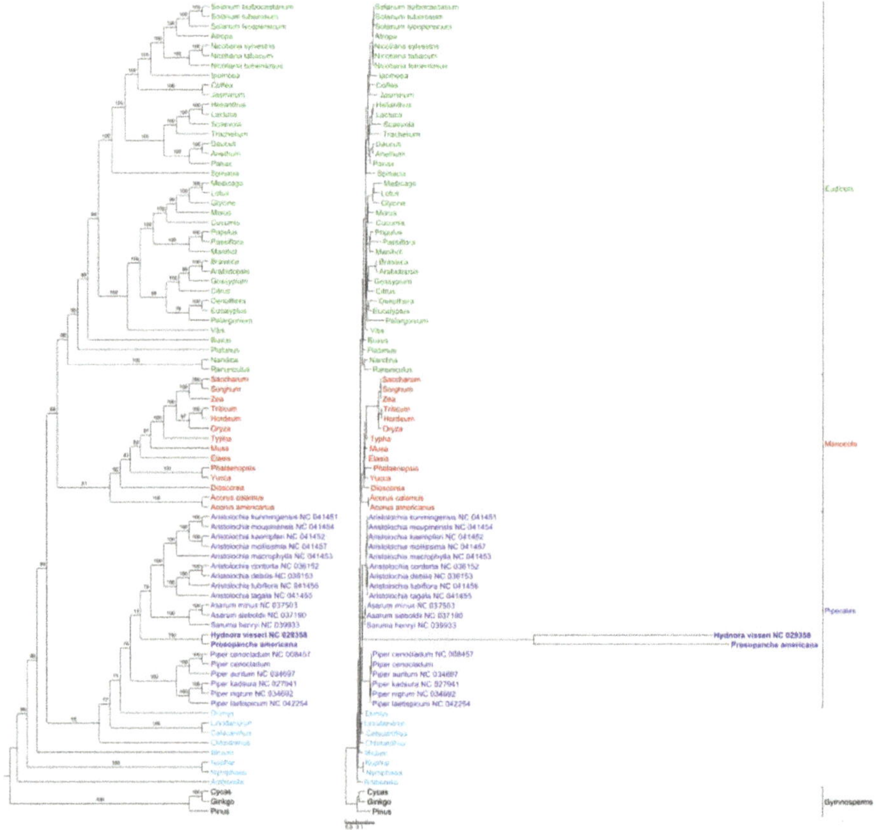

Figure 6. High mutational rate within Hydnoraceae relative to other Piperales and diverse photosynthetic angiosperms. Phylogenetic maximum likelihood tree reconstruction as cladogram (left) and phylogram (right) estimated using RAxML with the GTR + I + G model and conducting rapid bootstrapping (1000 replicates) recovers *Hydnora* and *Prosopanche* as sister genera and Hydnoraceae together as sister to a sister group of Aristolochioideae and Asaroideae (both Aristolochiaceae) within the Piperales. Bootstrap support values are displayed above the nodes. The scale bar shows the number of substitutions per site. Ingroup taxa are color coded, with eudicots being green, monocots red, Piperales dark blue and remaining Magnoliids and ANITA grade taxa light blue. Taxa refer to the dataset of Jansen et al. [42] if not otherwise indicated by GenBank number.

Phylogenetic tree reconstruction based on a gene set reduced to the genes that are present in Hydnoraceae, as well as on amino acid alignments of the complete and reduced gene set recovered relationships with generally lower support that are different from widely accepted topologies (Figures S2–S4).

3. Discussion

Phylogenetic tree reconstruction using maximum likelihood confidently verifies *Hydnora* and *Prosopanche* as sister genera (100 bootstrap support) and shows Hydnoraceae monophyletic, placed within Piperales. The placement of Hydnoraceae as sister to Aristolochiaceae in its widest circumscription (i.e., including Aristolochioideae and Asaroideae, Lactoridaceae not sampled) receives, however, no support (bootstrap value = 13) and might be a result of the long branches as well as missing data (few genes present in Hydnoraceae compared to other angiosperms). The inclusion of Hydnoraceae within Piperales is congruent with previous analyses. However, the positioning within the order differs between multiple different datasets and analyses. Naumann et al. [29] place Hydnoraceae as sister to *Aristolochia* and *Thottea* (Aristolochioideae) with *Lactoris* (Lactoridaceae) being sister to this group (Asaroideae not sampled). Nickrent et al. (six gene analysis) [33] reconstructed Hydnoraceae as sister to *Lactoris* (Lactoridaceae) and this group as sister to *Aristolochia*, but missing Asaroideae in the sampling. Massoni et al. [43] (12 gene analysis) recover Hydnoraceae as sister to *Lactoris* and Aristolochioideae with moderate support and this group sister to Asaroideae.

The plastome of the holoparasite *Prosopanche americana* was found on only one of the 372 scaffolds taken into consideration based on the BLAST for pt features with low stringency settings. The exclusion of the other scaffolds was due to a combination of coverage and quality of the BLAST hit. Careful analysis showed that many of the hits, often only a few bp, were for features that are similar between the different plant genomes (e.g., tRNA, rRNA), especially between the mitochondrial genome and the plastome. An additional BLAST with mitochondrial features (data not shown) revealed overlapping with the pt BLAST results. As a result, the plastome of *Prosopanche americana* solely consists of scaffold 424 and shows a high similarity to the plastome of the sister genus *Hydnora* with respect to genome size, gene content and order as well as levels of AT richness. Both Hydnoraceae plastomes contain the identical gene set with the exception of two additional tRNA genes in *Prosopanche* (*trn*W, *trn*Y). All ribosomal RNAs and protein coding genes are putatively functional (based on feature continuous reading frames), with *rps*7 being the exception for *Hydnora visseri* and *rps*19 the exception for *Prosopanche americana*. The plastome of *Aristolochia*, the closest available, photoautotrophic relative, is about six times the size of Hydnoraceae and contains over six times as many genes with the complete set of plastid-encoded photosynthetic genes. These findings make Hydnoraceae ideal candidates to visualize and compare the degree of reduction and gene loss in a parasitic lineage that is close to a potential endpoint, the potential complete plastome loss [14]. Orobanchaceae, a relatively young parasitic lineage [29], display the entire range of parasitic lifestyles [19] and allow for the visualization of most of the proposed stages of plastome breakdown [5,16,17]. However, plastomes of Orobanchaceae are not nearly as reduced in size or gene content as some mycoheterotrophes and a few parasitic plants, such as *Pylostyles* [12,21], *Balanophora* [22] or Hydnoraceae. The quadripartite plastome structure with two single copy regions and inverted repeats is common between most photoautotrophic plants and also retained in the majority of the parasitic lineages. *Hydnora* [10] still possesses an IR, though only containing one complete gene (*trn*I-CAU) and two pseudogenes. With a length of 1466 bp, it is much smaller than the one of *Aristolochia contorta* (25,459 bp) [34] containing 17 complete genes and one pseudogene. *Prosopanche americana* also contains a gene duplication (*trn*I-CAU), but as both copies appear in the same orientation, this is not to be considered an inverted repeat. The fact that the duplicated gene as well as the partial gene (*rpl*2) on the repeats can be found in the inverted repeat of *Aristolochia* as well as the extremely reduced IR of *Hydnora visseri* leads to the conclusion that these duplicated regions are remnants of an inverted repeat and potentially have changed orientation during a rearrangement process that occured along with genome reduction in *Prosopanche americana*. The coverage in both duplicated regions in *Prosopanche* is also not significantly higher than in the rest of the plastome, something usually common for inverted repeats. Additionally, the sequencing reads representing both copies are not assembled together as one sequence, which is the case for inverted repeats due to their identical sequence. The loss of one IR copy in *Prosopanche* could also be associated with the larger of the two inversions ('*trn*I-*rpl*2-*rps*2') when comparing the gene order of the two

Hydnoraceae genera. As these genes usually can be found in the inverted repeats, the loss of a different copy in *Prosopanche* compared to *Hydnora* could result in this apparent inverted gene order, as well as possible recombination events, for example, flip-flop recombination resulting in two plastome isoforms [44] before the IR loss in *Prosopanche*. The position of the second copy of the 195 bp duplication in the latter, nested in between the ribosomal RNA genes is noteworthy, as those are usually part of the inverted repeat. Therefore, the repeat position and the inverted region in comparison to *H. visseri* could potentially be a consequence of the IR loss, which is shown to lead to an increase in genome rearrangements. The occurrence of direct repeats instead of inverted repeats, although larger and containing more genes, is also known for species of the photoautotrophic genus *Selaginella* [45,46] and it is hypothesized that their recombination and gene conversion are not inhibited by the orientation of the direct repeat [47].

All protein coding genes present in the plastome of *Prosopanche americana* have a feature continuous reading frame, making them very likely functional (potential exception *rps*19) although evidence from transcriptome is currently unavailable. Functionality through structural analysis of the six tRNAs, as tested with tRNAscan-SE [38,39] is likely given for five of them. The sequence for *trn*W-CCA (Trp) is predicted by tRNAscan to have an ACA anticodon instead of a CCA anticodon. No tRNA with an ACA anticodon is known from Piperales plastomes but was found, for example, in Asterales (e.g., *Pogostemon cablin* MF287373 [48]). Sequence similarity strongly suggests that said Asterales tRNA with ACA anticodon is known as the intron-containing *trn*V-UAC in Piperales. Alignments of neither Asterales tRNA nor the *trn*V from Piperales did result in reasonable alignments.

Evaluation of nucleotide compositions of the shared protein coding genes between Hydnoraceae and *Aristolochia* visualizes the proportional increase in A and T nucleotides, also known from other plastomes of parasitic lineages [5,22]. *Hydnora* and *Prosopanche* showed a drastic decrease in G and C nucleotides across the whole plastome compared to the photoautotrophic *Aristolochia* with *Prosopanche* having slightly lower values than the sister genus *Hydnora*. Despite the drastically low proportion of G and C nucleotides in the Hydnoraceae plastomes compared to most photoautotrophic species, the record setting AT richness is known for holoparasitic *Balanophora* with most protein coding genes consisting to more than 90% of A and T [22]. For protein coding genes, mutations to the favored nucleotides predominantly occur on the third, variable codon position which has also been observed for other parasitic lineages such as Orobanchaceae [3]. Out of the three available stop codons (i.e., TAG, TGA and TAA) only two (TAG and TAA) are found in the protein coding genes of *Prosopanche*. The predominantly used stop codon found was TAA, highlighting another specific case of A and T nucleotide preference, whereas TAG was only found twice (*acc*D, *rpl*16) as stop codon. Similar cases of favoring certain stop codons or exclusively using a single one have been described, e.g., for the parasitic lineage *Cytinus* (Cytinaceae) [11] and *Balanophora* (Balanophoraceae) [22].

A comparison of the proportional amount that protein coding regions, ribosomal RNAs and IGRs contribute to the whole plastome highlights the many macro level changes Hydnoraceae plastomes experienced in comparison to the photoautotropic *Aristolochia*. The parasitic genomes have lost all plastid encoded genes related to the photosynthesis apparatus and their intergenic regions show a much more drastic decrease in length compared to *Aristolochia*, resulting in a proportional increase of space that the remaining protein coding genes and ribosomal RNAs take up in the plastome.

The comparisons of the two Hydnoraceae plastomes and the highlighted common set and order of genes between the genomes that have evolved independently since the split ~54 MYA lead to two possible scenarios. Either a similarly reduced plastome existed already in the MRCA of *Hydnora* and *Prosopanche* and did not experience many changes with respect to gene content and order since or the observed similarities are a result of two independent evolutionary pathways leading to the same endpoint. Although the first hypothesis would be more parsimonious, a larger sampling within Hydnoraceae would be needed to test these hypotheses. However, the second hypothesis would support the existence of taxon dependent essential gene sets for heterotrophic plants [3,49].

4. Materials and Methods

4.1. Plant Material, DNA Extraction, Library Preparation and Sequencing

Plant material of *Prosopanche americana* was collected near Chancaní (Córdoba, Argentina). A herbarium voucher with number AAC5681 was placed at Museo Botánico de Córdoba (CORD). Fresh tepal material was sliced and air dried for 24 hours and subsequently stored in silica gel. DNA extraction was done using the DNeasy Plant Maxi Kit (Qiagen, Venlo, Netherlands). Molecular weight of isolated DNA as well as quality was tested using gel electrophoresis, and NanoDrop 2000 (Thermo Scientific, Waltham, MA, U.S.). Illumina TruSeq DNA PCR-free library preparation was done using 2.2 µg genomic DNA with an insert size of 650 bp. The library was sequenced using a partial lane of an Illumina NextSeq High Output with 300 cycles creating 37,324 million reads. Both, the library preparation and library sequencing were performed at the DRESDEN concept Genome Center of TU Dresden (https://genomecenter.tu-dresden.de).

4.2. Assembly and Identification of Scaffolds with Plastid Information

Using the de novo assembly function, raw data were assembled with CLC Workbench [35] allowing for automatic word and bubble size. The assembly resulted in a total of 8,342,403 scaffolds to which the reads were mapped back to obtain coverage information and to be able to evaluate the assembly in specific regions of interest. The assembly was blasted (BLASTn, e-value e1-10) against a database containing 45 angiosperm plastid reference genomes from GenBank. Contigs with hits to plastid genomes were extracted and used together with the readmapping information for the creation of a stoichiometry plot in RStudio [50] allowing for highlighting of contigs with hits to plastid gene features followed by a directed and more efficient search for contigs potentially belonging to the plastid genome. The latter contigs were selected for further investigation.

4.3. Gene Annotation and Circularization

To check for and identify genes the scaffolds were uploaded and analyzed using DOGMA [36] at low stringencies (percent identity cutoff for protein coding genes and RNAs = 25, E-value = 1e-5) and additionally using MFannot [37]. Additionally, alignment tools such as "Map to reference" and LASTZ version 7.0.2 [51] were used as a plug in within Geneious [52] and applied to the scaffolds identified in the initial BLASTn search and the stoichiometry plot. In addition, the published plastome of *Hydnora visseri* (NC_029358) [10] was used to identify and annotate genes due to potentially higher sequence similarity between the sister genera. Gene boundaries were identified, then genes were annotated using Geneious [52] by aligning respective genes of *Hydnora visseri* and closely related photoautotrophic species of Piperales and in case of protein coding, genes feature continuous reading frames were checked manually. As a proxy of functionality of tRNA genes, in the absence of transcriptome data, the respective sequences were used as input for tRNAscan-SE [38,39] to predict their 2D structure. Criteria used were a cloverleaf structure with three leaves and that the predicted anticodon through tRNAscan found at the anticodon arm matches the anticodon predicted through sequence similarity. The full output is available as supplementary material (Table S1).

The *Prosopanche americana* plastome was finally circularized using the "Circular Sequence" tool in Geneious [52] and drawn using OGDRAW [53]. The correctness of this circular connection was verified using PCR followed by Sanger sequencing (forward and reverse) with a forward (ProAm-rps2F: AACTAAATTACAAGCCATTGATA) and a reverse primer (ProAm-rps14R: TCCTAGAGGTTATTATCGTTAT) derived from the available flanking regions.

4.4. Plastome Comparisons

The plastome of *Prosopanche americana* (GenBank accession number MT075717) was compared with the plastome of its sister genus *Hydnora* (*Hydnora visseri*, NC_029358) [10] as well as a plastome of its close autotrophic relative *Aristolochia* (*Aristolochia contorta*, NC_036152) [34] in terms of multiple parameters such as length, gene content and order, GC content and plastome structure with Geneious [40]. A dotplot (Score matrix: exact, window size: 100, threshold: 200), implemented in Geneious, was used to visualize differences between the *Prosopanche* and *Hydnora* plastome nucleotide sequences. For the protein coding genes shared between the three species, the rates of the four nucleotides were determined using CodonW (version 1.4.2) [41].

4.5. Phylogenetic Placement of Prosopanche Americana

Protein coding and rRNA genes of *Prosopanche americana* and *Hydnora visseri* were added and aligned to the 81 angiosperm-wide alignments published by Jansen et al. [42]. Additionally, an alignment for the *acc*D gene was created using data from GenBank. The sampling was then improved by adding all complete plastid genomes of Piperales accessions available in GenBank (https://www.ncbi.nlm.nih.gov/), increasing the data set to 83 taxa. After automatic alignment with MAFFT (version 7.450) [54,55], individual alignments were inspected and improved by eye using AliView (version 1.20) [56] and concatenated in Geneious [40]. Single gene alignments have been uploaded to TreeBASE (http://purl.org/phylo/treebase/phylows/study/TB2:S25836). RAxML analysis [57] using the GTR+I+G model with 1000 bootstrap replications was carried out on the concatenated data set using the CIPRES Science Gateway [58] after calculating the optimal substitution model using jModelTest (version 2.1.7) [59]. Phylogenetic trees were rooted using the three Gymnosperms (*Cycas, Ginkgo, Pinus*) as outgroup. The output was visualized in TreeGraph 2 [60]. In addition to the 82 gene analysis, we used a gene set reduced to the ones that are present in the Hydnoraceae (Figure S2) and also did RAxML analysis based on amino acid alignments (translated with Geneious [40]), both on the complete (Figure S3) and reduced gene set (Figure S4).

Supplementary Materials: The following are available online at http://www.mdpi.com/2223-7747/9/3/306/s1, Figure S1: Secondary structure of *Prosopanche americana* tRNAs. The secondary structure of the five unique *Prosopanche* tRNAs as predicted with tRNAscan-SE [38,39] shows the cloverleaf structure with three leaves for all but *trn*Y-GUA, which has an additional, variable arm. The predicted anticodon (boxes) matches the one predicted by DOGMA and through sequence similarity in all but the case of *trn*W-CCA, where tRNAscan predicts an ACA anticodon, Table S1: Anticodon prediction versus isotype model prediction visualizes tRNA sequence divergence in the *Prosopanche americana* plastome. Tabular tRNAscan-SE [38,39] output for the annotated *P. americana* tRNAs shows the anticodon predicted isotype, predicted anticodon and consistency of the anticodon versus isotype model predictions. Isotype scores are displayed with the top score being red, the 2nd highest score orange and the third highest score yellow, Table S2: Hydnoraceae protein coding genes show increase of A and T nucleotides at third codon position. The nucleotide usage as well as the GC content of the Hydnoraceae and *Aristolochia* protein coding genes as predicted by CodonW [41] highlights higher overall A and T content of the holoparasitic protein coding genes and shows the drastic differences at the third, variable codon position compared to photoautotrophic relative *Aristolochia contorta* [34], Figure S2: Angiosperm phylogeny based on a reduced plastid gene set. Phylogenetic maximum likelihood tree reconstruction as cladogram (left) and phylogram (right) based on a plastid marker set reduced to the genes present in the Hydnoraceae, estimated with RAxML using the GTR + I + G model and conducting rapid bootstrapping (1000 replicates) recovers *Hydnora* and *Prosopanche* as sister genera and Hydnoraceae together as sister to all other Piperales. Monocots are recovered as sister to eudicots and a group of Piperales together with *Drimys, Calycanthus* and *Liriodendron*. Bootstrap support values are displayed above the nodes. The scale bar shows the number of substitutions per site. Ingroup taxa are color coded, with eudicots being green, monocots red, Piperales dark blue and remaining Magnoliids and ANITA grade taxa light blue. Taxa refer to the dataset of Jansen et al. [42] if not otherwise indicated by GenBank number, Figure S3: Phylogenetic tree reconstruction based on amino acid alignments of plastid markers places Hydnoraceae as sister to all other angiosperms. Phylogenetic maximum likelihood tree reconstruction as cladogram (left) and phylogram (right) based on a translated and concatenated set of 78 plastid protein coding genes using RAxML and conducting rapid bootstrapping (1000 replicates) recovers *Hydnora* and *Prosopanche* as sister genera and Hydnoraceae together as sister to all other Angiosperms. Bootstrap support values are displayed above the nodes. The scale bar shows the number of substitutions per site. Ingroup taxa are color coded, with eudicots being green, monocots red, Piperales dark blue and remaining Magnoliids and ANITA grade taxa light blue. Taxa refer to the dataset of Jansen et al. [42] if not otherwise indicated by GenBank number, Figure S4 Phylogenetic tree reconstruction based on a reduced set of amino acid plastid markers places Hydnoraceae within eudicots. Phylogenetic maximum likelihood tree reconstruction as cladogram (left) and phylogram (right) based on a translated and concatenated set of plastid protein coding genes reduced to genes present in Hydnoraceae using RAxML and conducting rapid bootstrapping (1000 replicates) recovers *Hydnora* and *Prosopanche* as sister genera and places Hydnoraceae within Eudicots. *Illicium* is recovered as sister to all other Angiosperms. Bootstrap support values are displayed above the nodes. The scale bar shows the number of substitutions per site. Ingroup taxa are color coded, with eudicots being green, monocots red, Piperales dark blue and remaining Magnoliids and ANITA grade taxa light blue. Taxa refer to the dataset of Jansen et al. [42] if not otherwise indicated by GenBank number.

Author Contributions: S.W. conceived the study; fieldwork was done by N.R., A.A.C., S.W.; data generation and analyses M.J., J.N.; writing of the first draft M.J., S.W; responsible for visualization of results M.J., J.N.; all authors reviewed and edited the draft and agreed to the published version of the manuscript. All authors have read and agreed to the published version of the manuscript.

Funding: This research received no external funding

Acknowledgments: We thank Christoph Neinhuis for constant support of our work on Hydnoraceae, Sebastian Müller for assistance with lab work, and Gesine Schäfer for her initial work on the plastid genome of *Prosopanche*. We also thank Eric K. Wafula and Claude W. dePamphilis for assistance with the assembly and feedback on the manuscript.

Conflicts of Interest: The authors declare no conflict of interest.

Appendix A

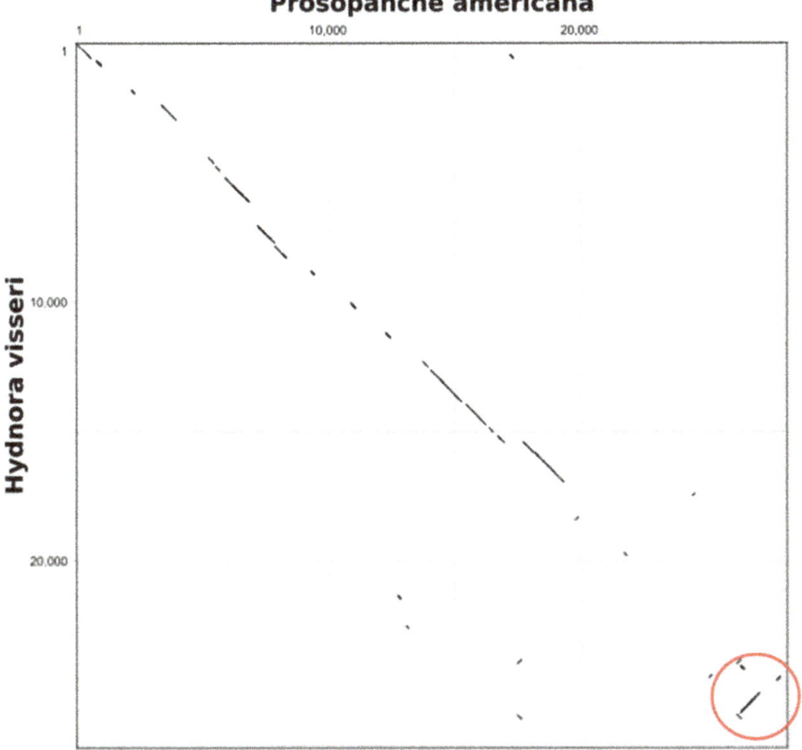

Figure A1. Dotplot comparison of the plastomes of *Prosopanche americana* and *Hydnora visseri* (NC_029358) highlights differences on nucleotide level and inversion (red circle). The diagonal lines indicate sequence similarity between the plastomes of *Hydnora* and *Prosopanche*, the gaps visualize areas of drastic nucleotide sequence differences. The red circle highlights the larger of the two inversions between the Hydnoraceae plastomes. The dotplot was created with Geneious (v 11.1.5) by using an exact score matrix, a window size of 100 and a threshold of 200.

References

1. Palmer, J.D. Comparative organization of chloroplast genomes. *Annu. Rev. Genet.* **1985**, *19*, 325–354. [CrossRef]
2. Ohyama, K.; Fukuzawa, H.; Kohchi, T.; Shirai, H.; Sano, T.; Sano, S.; Umesono, K.; Shiki, Y.; Takeuchi, Y.; Chang, Z.; et al. Chloroplast gene organization deduced from complete sequence of liverwort Marchantia polymorpha chloroplast DNA. *Nature* **1986**, *322*, 572. [CrossRef]
3. Wicke, S.; Müller, K.F.; de Pamphilis, C.W.; Quandt, D.; Wickett, N.J.; Zhang, Y.; Renner, S.S.; Schneeweiss, G.M. Mechanisms of functional and physical genome reduction in photosynthetic and nonphotosynthetic parasitic plants of the broomrape family. *Plant Cell* **2013**, *25*, 3711–3725. [CrossRef] [PubMed]
4. Krause, K. Piecing together the puzzle of parasitic plant plastome evolution. *Planta* **2011**, *234*, 647. [CrossRef] [PubMed]
5. Wicke, S.; Naumann, J. Molecular evolution of plastid genomes in parasitic flowering plants. In *Advances in Botanical Research*; Academic Press: Cambridge, MA, USA, 2018; Volume 85, pp. 315–347.
6. Barkman, T.J.; McNeal, J.R.; Lim, S.H.; Coat, G.; Croom, H.B.; Young, N.D.; Claude, W.D. Mitochondrial DNA suggests at least 11 origins of parasitism in angiosperms and reveals genomic chimerism in parasitic plants. *BMC Evol. Biol.* **2007**, *7*, 248. [CrossRef]
7. Schneider, A.C.; Braukmann, T.; Banerjee, A.; Stefanovic, S. Convergent plastome evolution and gene loss in holoparasitic Lennoaceae (Boraginales). *Genome Biol. Evol.* **2018**, *10*, 2663–2670. [CrossRef]
8. Wu, C.S.; Wang, T.J.; Wu, C.W.; Wang, Y.N.; Chaw, S.M. Plastome evolution in the sole hemiparasitic genus laurel dodder (Cassytha) and insights into the plastid phylogenomics of Lauraceae. *Genome Biol. Evol.* **2017**, *9*, 2604–2614. [CrossRef]
9. Bellot, S.; Cusimano, N.; Luo, S.; Sun, G.; Zarre, S.; Gröger, A.; Temsch, E.; Renner, S.S. Assembled plastid and mitochondrial genomes, as well as nuclear genes, place the parasite family Cynomoriaceae in the Saxifragales. *Genome Biol. Evol.* **2016**, *8*, 2214–2230. [CrossRef]
10. Naumann, J.; Der, J.P.; Wafula, E.K.; Jones, S.S.; Wagner, S.T.; Honaas, L.A.; Ralph, P.E.; Bolin, J.F.; Maass, E.; Neinhuis, C.; et al. Detecting and characterizing the highly divergent plastid genome of the nonphotosynthetic parasitic plant Hydnora visseri (Hydnoraceae). *Genome Biol. Evol.* **2016**, *8*, 345–363. [CrossRef]
11. Roquet, C.; Coissac, É.; Cruaud, C.; Boleda, M.; Boyer, F.; Alberti, A.; Gielly, L.; Taberlet, P.; Thuiller, W.; Van Es, J.; et al. Understanding the evolution of holoparasitic plants: The complete plastid genome of the holoparasite Cytinus hypocistis (Cytinaceae). *Ann. Bot.* **2016**, *118*, 885–896. [CrossRef]
12. Bellot, S.; Renner, S.S. The plastomes of two species in the endoparasite genus Pilostyles (Apodanthaceae) each retain just five or six possibly functional genes. *Genome Biol. Evol.* **2015**, *8*, 189–201. [CrossRef] [PubMed]
13. Petersen, G.; Cuenca, A.; Seberg, O. Plastome evolution in hemiparasitic mistletoes. *Genome Biol. Evol.* **2015**, *7*, 2520–2532. [CrossRef] [PubMed]
14. Molina, J.; Hazzouri, K.M.; Nickrent, D.; Geisler, M.; Meyer, R.S.; Pentony, M.M.; Flowers, J.M.; Pelser, P.; Barcelona, J.; Inovejas, S.A.; et al. Possible loss of the chloroplast genome in the parasitic flowering plant Rafflesia lagascae (Rafflesiaceae). *Mol. Biol. Evol.* **2014**, *31*, 793–803. [CrossRef] [PubMed]
15. Funk, H.T.; Berg, S.; Krupinska, K.; Maier, U.G.; Krause, K. Complete DNA sequences of the plastid genomes of two parasitic flowering plant species, Cuscuta reflexa and Cuscuta gronovii. *BMC Plant Biol.* **2007**, *7*, 45. [CrossRef] [PubMed]
16. Graham, S.W.; Lam, V.K.; Merckx, V.S. Plastomes on the edge: The evolutionary breakdown of mycoheterotroph plastid genomes. *New Phytol.* **2017**, *214*, 48–55. [CrossRef] [PubMed]
17. Barrett, C.F.; Davis, J.I. The plastid genome of the mycoheterotrophic Corallorhiza striata (Orchidaceae) is in the relatively early stages of degradation. *Am. J. Bot.* **2012**, *99*, 1513–1523. [CrossRef]
18. Wicke, S.; Müller, K.F.; dePamphilis, C.W.; Quandt, D.; Bellot, S.; Schneeweiss, G.M. Mechanistic model of evolutionary rate variation en route to a nonphotosynthetic lifestyle in plants. *Proc. Natl. Acad. Sci. USA* **2016**, *113*, 9045–9050. [CrossRef]
19. Wicke, S. Genomic evolution in Orobanchaceae. In *Parasitic Orobanchaceae*; Springer: Berlin/Heidelberg, Germany, 2013; pp. 267–286.
20. Westwood, J.H.; Yoder, J.I.; Timko, M.P.; dePamphilis, C.W. The evolution of parasitism in plants. *Trends Plant Sci.* **2010**, *15*, 227–235. [CrossRef]

21. Arias-Agudelo, L.M.; González, F.; Isaza, J.P.; Alzate, J.F.; Pabón-Mora, N. Plastome reduction and gene content in New World Pilostyles (Apodanthaceae) unveils high similarities to African and Australian congeners. *Mol. Phylogenet. Evol.* **2019**, *135*, 193–202. [CrossRef]

22. Su, H.J.; Barkman, T.J.; Hao, W.; Jones, S.S.; Naumann, J.; Skippington, E.; Wafula, E.K.; Hu, J.M.; Palmer, J.D.; dePamphilis, C.W. Novel genetic code and record-setting AT-richness in the highly reduced plastid genome of the holoparasitic plant Balanophora. *Proc. Natl. Acad. Sci. USA* **2019**, *116*, 934–943. [CrossRef]

23. Lavin, M.; Doyle, J.J.; Palmer, J.D. Evolutionary significance of the loss of the chloroplast-DNA inverted repeat in the Leguminosae subfamily Papilionoideae. *Evolution* **1990**, *44*, 390–402. [CrossRef] [PubMed]

24. Palmer, J.D.; Osorio, B.; Aldrich, J.; Thompson, W.F. Chloroplast DNA evolution among legumes: Loss of a large inverted repeat occurred prior to other sequence rearrangements. *Curr. Genet.* **1987**, *11*, 275–286. [CrossRef]

25. Wicke, S.; Schneeweiss, G.M.; Depamphilis, C.W.; Müller, K.F.; Quandt, D. The evolution of the plastid chromosome in land plants: Gene content, gene order, gene function. *Plant Mol. Biol.* **2011**, *76*, 273–297. [CrossRef] [PubMed]

26. Zhu, A.; Guo, W.; Gupta, S.; Fan, W.; Mower, J.P. Evolutionary dynamics of the plastid inverted repeat: The effects of expansion, contraction, and loss on substitution rates. *New Phytol.* **2016**, *209*, 1747–1756. [CrossRef] [PubMed]

27. Maréchal, A.; Brisson, N. Recombination and the maintenance of plant organelle genome stability. *New Phytol.* **2010**, *186*, 299–317. [CrossRef] [PubMed]

28. Palmer, J.D.; Thompson, W.F. Chloroplast DNA rearrangements are more frequent when a large inverted repeat sequence is lost. *Cell* **1982**, *29*, 537–550. [CrossRef]

29. Naumann, J.; Salomo, K.; Der, J.P.; Wafula, E.K.; Bolin, J.F.; Maass, E.; Frenzke, L.; Samain, M.S.; Neinhuis, C.; dePamphilis, C.W.; et al. Single-copy nuclear genes place haustorial Hydnoraceae within Piperales and reveal a Cretaceous origin of multiple parasitic angiosperm lineages. *PLoS ONE* **2013**, *8*, e79204. [CrossRef]

30. Bolin, J.F.; Lupton, D.; Musselman, L.J. Hydnora arabica (Aristolochiaceae), a new species from the Arabian Peninsula and a key to Hydnora. *Phytotaxa* **2018**, *338*, 99. [CrossRef]

31. Funez, L.A.; Ribeiro-Nardes, W.; Kossmann, T.; Peroni, N.; Drechsler-Santos, E.R. Prosopanche demogorgoni: A new species of Prosopanche (Aristolochiaceae: Hydnoroideae) from southern Brazil. *Phytotaxa* **2019**, *422*, 93–100. [CrossRef]

32. Musselman, L.J.; Visser, J.H. Taxonomy and natural history of Hydnora (Hydnoraceae). *Aliso J. Syst. Evol. Bot.* **1989**, *12*, 317–326. [CrossRef]

33. Nickrent, D.L.; Blarer, A.; Qiu, Y.L.; Soltis, D.E.; Soltis, P.S.; Zanis, M. Molecular data place Hydnoraceae with Aristolochiaceae. *Am. J. Bot.* **2002**, *89*, 1809–1817. [CrossRef] [PubMed]

34. Zhou, J.; Chen, X.; Cui, Y.; Sun, W.; Li, Y.; Wang, Y.; Song, J.; Yao, H. Molecular structure and phylogenetic analyses of complete chloroplast genomes of two Aristolochia medicinal species. *Int. J. Mol. Sci.* **2017**, *18*, 1839. [CrossRef] [PubMed]

35. CLC. Genomics Workbench. Available online: https://www.qiagenbioinformatics.com (accessed on 15 February 2020).

36. Wyman, S.K.; Jansen, R.K.; Boore, J.L. Automatic annotation of organellar genomes with DOGMA. *Bioinformatics* **2014**, *20*, 3252–3255. [CrossRef] [PubMed]

37. Beck, N.; Lang, B. *MFannot, Organelle Genome Annotation Websever*; Université de Montréal: Montréal, QC, Canada, 2010.

38. Lowe, T.M.; Chan, P.P. tRNAscan-SE On-line: Search and Contextual Analysis of Transfer RNA Genes. *Nucleic Acids Res.* **2016**, *44*, W54–W57. [CrossRef] [PubMed]

39. Lowe, T.M.; Eddy, S.R. tRNAscan-SE: A program for improved detection of transfer RNA genes in genomic sequence. *Nucleic Acids Res.* **1997**, *25*, 955–964. [CrossRef]

40. Geneious 11.1.5. Available online: https://www.geneious.com (accessed on 22 January 2020).

41. Peden, J. CodonW Version 1.4. 2. 2005. Available online: http://codonw.sourceforge.net/ (accessed on 15 February 2020).

42. Jansen, R.K.; Cai, Z.; Raubeson, L.A.; Daniell, H.; Depamphilis, C.W.; Leebens-Mack, J.; Müller, K.F.; Guisinger-Bellian, M.; Haberle, R.C.; Hansen, A.K.; et al. Analysis of 81 genes from 64 plastid genomes resolves relationships in angiosperms and identifies genome-scale evolutionary patterns. *Proc. Natl. Acad. Sci. USA* **2007**, *104*, 19369–19374. [CrossRef]

43. Massoni, J.; Forest, F.; Sauquet, H. Increased sampling of both genes and taxa improves resolution of phylogenetic relationships within Magnoliidae, a large and early-diverging clade of angiosperms. *Mol. Phylogenet. Evol.* **2014**, *70*, 84–93. [CrossRef]

44. Palmer, J.D. Chloroplast DNA exists in two orientations. *Nature* **1983**, *301*, 92–93. [CrossRef]

45. Zhang, H.R.; Zhang, X.C.; Xiang, Q.P. Directed repeats co-occur with few short-dispersed repeats in plastid genome of a Spikemoss, Selaginella vardei (Selaginellaceae, Lycopodiopsida). *BMC Genom.* **2019**, *20*, 484.

46. Xu, Z.; Xin, T.; Bartels, D.; Li, Y.; Gu, W.; Yao, H.; Liu, S.; Yu, H.; Pu, X.; Zhou, J.; et al. Genome analysis of the ancient tracheophyte Selaginella tamariscina reveals evolutionary features relevant to the acquisition of desiccation tolerance. *Mol. Plant* **2018**, *11*, 983–994. [CrossRef]

47. Mower, J.P.; Ma, P.F.; Grewe, F.; Taylor, A.; Michael, T.P.; VanBuren, R.; Qiu, Y.L. Lycophyte plastid genomics: Extreme variation in GC, gene and intron content and multiple inversions between a direct and inverted orientation of the rRNA repeat. *New Phytol.* **2019**, *222*, 1061–1075. [CrossRef]

48. Zhang, C.; Liu, T.; Yuan, X.; Huang, H.; Yao, G.; Mo, X.; Xue, X.; Yan, H. The plastid genome and its implications in barcoding specific-chemotypes of the medicinal herb Pogostemon cablin in China. *PLoS ONE* **2019**, *14*, e0215512. [CrossRef] [PubMed]

49. Schelkunov, M.I.; Shtratnikova, V.Y.; Nuraliev, M.S.; Selosse, M.A.; Penin, A.A.; Logacheva, M.D. Exploring the limits for reduction of plastid genomes: A case study of the mycoheterotrophic orchids Epipogium aphyllum and Epipogium roseum. *Genome Biol. Evol.* **2015**, *7*, 1179–1191. [CrossRef] [PubMed]

50. RStudio Team. *RStudio: Integrated Development for R*; RStudio, Inc.: Boston, MA, USA, 2016; Available online: https://rstudio.com/ (accessed on 15 February 2020).

51. Harris, R.S. Improved Pairwise Alignmnet of Genomic DNA. Ph.D. Thesis, Penn State University, State College, PA, USA, 2007.

52. Geneious 8.1. Available online: https://www.geneious.com (accessed on 22 January 2020).

53. Greiner, S.; Lehwark, P.; Bock, R. OrganellarGenomeDRAW (OGDRAW) version 1.3. 1: Expanded toolkit for the graphical visualization of organellar genomes. *Nucleic Acids Res.* **2019**, *47*, W59–W64. [CrossRef] [PubMed]

54. Katoh, K.; Standley, D.M. MAFFT multiple sequence alignment software version 7: Improvements in performance and usability. *Mol. Biol. Evol.* **2013**, *30*, 772–780. [CrossRef]

55. Katoh, K.; Misawa, K.; Kuma, K.I.; Miyata, T. MAFFT: A novel method for rapid multiple sequence alignment based on fast Fourier transform. *Nucleic Acids Res.* **2002**, *30*, 3059–3066. [CrossRef]

56. Larsson, A. AliView: A fast and lightweight alignment viewer and editor for large datasets. *Bioinformatics* **2014**, *30*, 3276–3278. [CrossRef]

57. Stamatakis, A. RAxML Version 8: A tool for Phylogenetic Analysis and Post-Analysis of Large Phylogenies. *Bioinformatics* **2014**, *30*, 1312–1313. [CrossRef]

58. Miller, M.A.; Pfeiffer, W.; Schwartz, T. Creating the CIPRES Science Gateway for inference of large phylogenetic trees. In Proceedings of the Gateway Computing Environments Workshop (GCE), New Orleans, LA, USA, 14 November 2010; pp. 1–8.

59. Darriba, D.; Taboada, G.L.; Doallo, R.; Posada, D. jModelTest 2: More models, new heuristics and parallel computing. *Nat. Methods* **2012**, *9*, 772. [CrossRef]

60. Stöver, B.C.; Müller, K.F. TreeGraph 2: Combining and visualizing evidence from different phylogenetic analyses. *BMC Bioinform.* **2010**, *11*, 7. [CrossRef]

MDPI

St. Alban-Anlage 66

4052 Basel

Switzerland

Tel. +41 61 683 77 34

Fax +41 61 302 89 18

www.mdpi.com

Plants Editorial Office

E-mail: plants@mdpi.com

www.mdpi.com/journal/plants